RELIABILITY AND MAINTENANCE ENGINEERING

RELIABILITY AND MAINTENANCE ENGINEERING

R.C. Mishra
Professor & Head, Mechanical Engineering Department
Azad Institute Engineering & Technology, Lucknow

Formerly Professor & Head, Dept. of Mechanical Engineering
and Mining Machinery Engineering
Indian School of Mines, Dhanbad

PUBLISHING FOR ONE WORLD

NEW AGE INTERNATIONAL (P) LIMITED, PUBLISHERS
New Delhi · Bangalore · Chennai · Cochin · Guwahati · Hyderabad
Jalandhar · Kolkata · Lucknow · Mumbai · Ranchi
Visit us at **www.newagepublishers.com**

Published by New Age International (P) Ltd., Publishers
First Edition: 2006
Reprint: 2015

BRANCHES

- **Bangalore** 37/10, 8th Cross (Near Hanuman Temple), Azad Nagar, Chamarajpet, Bangalore-560 018 Tel.: (080) 26756823, Telefax: 26756820, **E-mail: bangalore@newagepublishers.com**
- **Chennai** 26, Damodaran Street, T. Nagar, Chennai-600 017, Tel.: (044) 24353401 Telefax: 24351463, **E-mail: chennai@newagepublishers.com**
- **Cochin** CC-39/1016, Carrier Station Road, Ernakulam South, Cochin-682 016 Tel.: (0484) 2377303, Telefax: 4051303, **E-mail: cochin@newagepublishers.com**
- **Guwahati** Hemsen Complex, Mohd. Shah Road, Paltan Bazar, Near Starline Hotel, Guwahati-781 008 Tel.: (0361) 2513881, Telefax: 2543669, **E-mail: guwahati@newagepublishers.com**
- **Hyderabad** 105, 1st Floor, Madhiray Kaveri Tower, 3-2-19, Azam Jahi Road, Near Kumar Theater Nimboliadda, Kachiguda, Hyderabad-500 027, Tel.: (040) 24652456, Telefax: 24652457 **E-mail: hyderabad@newagepublishers.com**
- **Kolkata** RDB Chambers (Formerly Lotus Cinema) 106A, 1st Floor, S.N. Banerjee Road, Kolkata-700 014 Tel.: (033) 22273773, Telefax: 22275247, **E-mail: kolkata@newagepublishers.com**
- **Lucknow** 16-A, Jopling Road, Lucknow-226 001, Tel.: (0522) 2209578, 4045297, Telefax: 2204098 **E-mail: lucknow@newagepublishers.com**
- **Mumbai** 142C, Victor House, Ground Floor, N.M. Joshi Marg, Lower Parel, Mumbai-400 013 Tel.: (022) 24927869, Telefax: 24915415, **E-mail: mumbai@newagepublishers.com**
- **New Delhi** 22, Golden House, Daryaganj, New Delhi-110 002, Tel.: (011) 23262368, 23262370 Telefax: 43551305, **E-mail: sales@newagepublishers.com**

ISBN: 978-81-224-1741-8

₹ 175.00

C-15-03-8336

Printed in India at Ajit Printing Press, Delhi.
Typeset at Innovative Processors, Delhi.

PUBLISHING GLOBALLY
NEW AGE INTERNATIONAL (P) LIMITED, PUBLISHERS
7/30 A, Daryaganj, New Delhi-110002
Visit us at **www.newagepublishers.com**

Preface

In the today's context the need and requirements of Maintenance and Reliability cannot be neglected, and draws proper attention of each one who are concerned with it. Reliability is the need of the hour, as everybody is cautious, while purchasing any item/product. Therefore its basic knowledge is essential for everybody. In this book attempt has been made to bring out reliability concept in a simple and understandable language, though many books on the subject matter are available. In the first few chapters basics of reliability have illustrated besides its application in actual conditions with examples. Effects of design and quality on reliability have also been discussed with its improvements and assessment. *The reliability testing has also been discussed.*

Maintenance is the most vital and indispensable jobs of any organization and its requirements have increased many folds with the growth of industrial sector. Such a necessity has brought about a major paradigm shift in the total approach to maintenance in the modern industry. The present text introduces the basics of this expanding subject to the undergraduate students of the engineering and provide them with a broad view of maintenance and reliability that would assist them in taking better managerial decisions wherever situation demands.

There are a few books available on the subject matter but this text bridges the gap between two topics i.e. Maintenance and Reliability. Under last few chapters, matter related to maintenance have been discussed which includes, maintenance concepts, maintenance practices, condition-based maintenance, application of computers in maintenance, and other related aspects of maintenance. The importance of spare parts, equipment and facilities for maintenance has also been included in the text.

The author acknowledges sincere thanks to his wife Smt. Urmila Mishra for extending moral support in bringing out this manuscript.

R. C. Mishra

Contents

1

Principles of Reliability

INTRODUCTION

The word reliability is associated with the civilization of mankind to compare one item/person with another. Trustworthy, dependable and consistent are the words, which can be used to give an indication of why the characteristic of reliability is valued. Reliability cannot be precisely measured with respect to human behaviour but can give an indication that a particular person is more reliable than the other. The judgment can be easier when it is associated with human behaviour or function. For example, the degree of punctuality of individuals for attending their work place or attending their work without failure can be used to measure their reliability in performing a particular function.

In case of man made products/systems the role of design engineer becomes more crucial to eliminate/minimize the large scale failures to avoid accidents in case of automotive vehicles etc. where risk for human life is involved. The knowledge about long-term properties of material and other devices helps in designing reliable products.

In fact, the characteristic of reliability is usually used to describe some function or task or in widest sense, it may be said to be a measure of performance. A person who completes his work in time is said to be more reliable than the other who does not. Now it is easy to state, that the concept of reliability is not only associated with human behaviour or activity but can also be applied to other inventions of the mankind, by directly measuring their performance or by knowing the failure rates of the equipment/systems. The growing awareness of reliability arises from the fact that there is a need for efficient, economic and continuous running of equipment/system in any organization for achieving the targeted production at a minimum cost to face the present competitive world. In World War II the need for reliability was felt because of failure of many military operations in spite of the best efforts from the users. After World War II a study was conducted which revealed the following facts:

(*i*) The electronic equipments/systems used by the Navy were operative for only 30% of its total available time because of frequent failures/maintenance problems.

(*ii*) Army equipments were either under repair/breakdown or commissioning for almost 60-75 % of its total time, which again created problems for a successful mission.

(*iii*) Air-force study conducted for about 5 years shown that the maintenance costs were very high and even sometimes many fold of the original cost of the equipments/systems due to frequent breakdowns.

The above facts may be for the reason that during war period the availability of equipments is of prime importance besides its cost. With the above information investigations were further carried out and a concept of reliability group/engineer came into existence. Since then efforts are continuing in this area to achieve the desired goals in industrial organizations. The concept of reliability in the minds of the people/engineers in general has changed their attitude towards the design and manufacturing processes as well. In other words, now the reliability is the concern of every one irrespective of design engineer or process engineer. Even in maintenance also the reliability has its own importance and can yield better results if taken care off.

Reliability Requirements for a System

These can be based on prototypes of a new technical system or based on practical experience for the system in use. This usually requires statistical data about similar objects/items assembled from similar components with expertise and statistics. Ideal conditions for the determination of reliability requirements exist when,

- The system/equipment output can be measured
- The system structure, operating characteristics etc. are well defined
- Necessary statistical data for all items/components are determined with satisfactory confidence.

SCOPE OF RELIABILITY

The scope of reliability can be visualized by the following facts in respect of any equipment or system.

- The working environment of the equipment/system.
- The need of safety aspects for men and material.
- Degree of uncertainty about the success of operation and its improvements in system/ equipment performance.
- Need for efficient, economic and continuous running of equipment/system without disturbances.
- A failure of an equipment/system raises the question in the minds of the people regarding its reliability and its further use.
- Improvement in the confidence of the working personnel particularly in the hazardous area because of safety reasons.

Any equipment/system manufactured is designed with certain objectives to meet its goals in terms of production/service. With passage of time or environmental conditions the equipment may not give the desired results over a period of time. This system may be treated as unreliable. Therefore a reliable equipment is the one which works for a given stipulated time period under given environmental conditions without interruptions. In general terms, if the failures are unpredicted and more in number, the equipment/system is said to be unreliable and its market value is reduced. On the other hand if the number of failures are less, then the equipment/system is reliable and its market value is high. The term risk is associated with frequent failures of equipment and to a large extent is a qualitative in nature and attempts are being made to assure its quantitative measure. It tries to combine the probability of something going wrong and consequences of the event. Now the problem is to assess these consequences. In certain cases it is easy and straightforward, whereas in some cases it is difficult to measure it. Keeping this in mind the risk can be defined as the product of its frequency and the magnitude. The frequency here means the failure of the equipment per unit time. These failures can occur due to various reasons depending upon the working conditions, operator skill and other such parameters.

To know the real scope of reliability it is required to highlight some technological systems, which need high reliability. The application of reliability becomes more dominant where risk factor is high and capital involvement is at stake with human life. The first in this class falls the aerospace systems mainly aircrafts. It is known that capital investment in case of aircrafts is very high and also human lives are at risk, the systems used must be highly reliable. The other area where high reliability is required is the nuclear power plant. If such plants are not available for power production, they incur heavy loss per hour as well as the goodwill of the concerning people. The radiations from such plants can cause havoc, due care has to be taken during the discharge of their by-products. Two main types of equipment failures normally affect the reliability of such systems. First of all the 'fail dangerous', and secondly the 'fail safe'. The fail dangerous will protect the system in case of any fault, whereas in fail safe mode the system will be shutdown in case of any fault. The above two systems must be highly reliable to avoid any mis-happening. Safety circuits can also be provided for the whole protective system. It is desired that the protective systems should shutdown the plant in case of serious fault where protection device fails.

The other highly reliable systems include chemical plants, process industries and electric supply systems where the failures can cause high revenue losses. Although the safety of human is not at stake in these cases but inconveniences are caused to the users. For example, running of electric trains can be badly affected by unreliable power supply.

OBJECTIVES OF RELIABILITY

During the design phase of any product/system, it is desired that the said system should meet the performance standards as expected within the specified constraints. These constraints include cost of equipment/product, environmental conditions and availability of material/parts etc. A system/equipment normally comprises of many units/components and their interconnection and integration should result in satisfactory performance. The number of components and units make a system complex and therefore system is dependent on complexity of the functioning of the units. It is further more difficult to achieve satisfactory performance from such system/equipment. Therefore, the objectives of reliability are many fold and include the following.

(*i*) Trouble free running of system/equipment
(*ii*) The adequate performance for a stated period of time
(*iii*) The equipment/system should work under the specified environmental conditions
(*iv*) Minimization of downtime of equipment/system and
(*v*) Maintainability of device/components

THE MEANING OF RELIABILITY

Any equipment/system after its manufacture or fabrication is tested under actual working conditions, and where actual conditions are not possible the concept of simulation is applied.

It is desired from the equipment/systems that they perform within the specified limits. If the product/system survives the tests, the reliability R of the equipment can be defined as the probability that the product continues to meet the given specifications over a given period of time subjected to environmental conditions. However, if the product does not meet the specifications as desired over a period of time it is considered as failed in its mission. The unreliability '*F*' of the product can be defined as the probability that the product fails to meet the specifications over a given time period under the given environmental conditions. The failures can occur due to many reasons known or unknown such as wear, mechanical fracture and chemical corrosion etc.

The reliability and unreliability are dependent on time. At time zero when the product/equipment is put into service its reliability is unity, after certain time interval it may be 0.5 and zero when the equipment, system fails. On the other hand unreliability increases with passage of time i.e. when the products tested and put into the service its unreliability is zero. With passage of time it may increase to 0.5 and finally to unity when the equipment/system fails. Since at any time '*t*' the product has either survived or failed, therefore the sum of reliability and unreliability must be unity i.e. the two events are complementary and can be expressed by the equation below:

$$R(t) + F(t) = 1 \quad ...(1.1)$$

Where (*t*) is the expected time period of equipment/system in use under the given environmental conditions?

Besides the quality of product, maintenances play an important role to improve the reliability. The high quality product may have high reliability, whereas reverse may not be true i.e. high reliability product may not be of high quality but it meets all specifications as desired for its operation.

Now we can establish relationship between quality and reliability. The reliability of a product is its ability to retain its quality during its usage. If the quality of the product is maintained with passage of time its reliability will also be maintained. The quality again can be kept intact with the use of better maintenance practices.

Definition of Reliability

The basic definition of reliability states that it is the probability of a device to perform its intended functions over a pre-specified time period under the given environmental conditions. Thus the important factors associated with the reliability are probability, intended functions, time period and the working conditions.

Since the reliability is expressed as probability, its value will vary from zero to unity thereby giving quantitative measure. In our daily life we compare one product over the other for its superiority in terms of either dependability or reliability. The degree of superiority can only be expressed in quantitative terms. The pre-mature failure of a product/system will help to establish the creditability of manufacturer for the superiority/quality when compared with other manufacturer for the same product. It is generally said that the product of A is superior to that of B based on the above statement. The ordering of the products can also be specified in quantitative terms as per their reliabilities.

The second aspect namely the intended functions can be expressed easily for some products such as electric/electronic devices, whereas it is difficult to specify the same when the product is complicated for example a turbine, which includes many components. For such devices the overall performance has to be examined in terms of reliability. Larger the system, the complications will increase many fold, because in a failed unit (turbine) some components will always be operative. But the failed unit will not allow the working of the equipment. Therefore, for reliability evaluation unit is to be considered as a whole, though made up of many components/parts.

The time period for which the reliability assessment has to be carried out is a complicated issue until and unless it is specified properly based on the past experience it may cause problems. For mechanical equipment it is easy to state the time limit since their failures are gradual in nature whereas, for electrical and electronic equipment it is really difficult since their failures are unpredictable and reasons for failures may be many.

For complete definition of reliability it is essential to know the following terms associated with it.

- *Total up time:* It is the period for which an equipment/system remains operational without any failure.
- *Mean time to failure:* It is the time period between the failures of equipment/system i.e. time of first failure to the next one.

Mean Failure Rate: The failures occurring over a time interval can be expressed as a mean failure rate. If there are N components in service for time T, then failure rate is N/T.

The total number of failures which have occurred or likely to occur in a particular equipment or system are of interest and could be used as some index of reliability.

For practical definition of reliability two cases can be considered where reliability calculations are independent of time.

Non-Repairable Items

In this case once the item fails it is to be replaced. Let us consider that there are N products placed in service with a test interval of T. The ith failure takes place at time T_i. The total up time with N failures is therefore $\sum_{i=1}^{N} T_i$, and the Mean Time to Failure ($MTTF$) is given by:

$$MTTF = \frac{1}{N}\sum_{i=1}^{N} T_i \qquad \text{...(1.2)}$$

The mean failure rate $\bar{\lambda}$ is expressed as:

$$= \frac{\text{Number of Failures}}{\text{Total up Time}}$$

$$\bar{\lambda} = \frac{N}{\sum_{i=1}^{N} T_i} \qquad \text{...(1.3)}$$

$MTTF$ can also be expressed in terms of reliability:

$$MTTF = \int_0^{\infty} R_t(dt) \qquad \text{...(1.4)}$$

Here ∞ means N being infinite

Repairable Items

In this case items/products can be repaired, which involves downtime of the product as it needs some time to repair. The downtime T_{Dj} is associated with j^{th} failure is the total time between the occurrence of the failure and the repaired item being put back into operation. The total downtime for N_F failures, therefore is $\sum_{J=1}^{N} T_{Dj}$ and Mean Down Time (*MDT*) is given by:

$$MDT = \frac{1}{N}\sum_{J=1}^{N_F} T_{DJ} \qquad \text{...(1.5)}$$

The total up time can be found out by subtracting total downtime from *NT* and is expressed as:

$$\text{Total Up Time} = NT - \sum_{J=1}^{N_F} T_{DJ} \qquad \text{...(1.6)}$$

$$= NT - N_F MDT$$

Where N_F is total number of failures.

The Mean Time Between Failure (*MTBF*) can be expressed as:

$$MTBF = \frac{NT - N_F MDT}{N_F}$$, and mean failure rate $\bar{\lambda}$ is given by,

$$\bar{\lambda} = \frac{N_F}{NT - N_F MDT} \qquad \text{...(1.8)}$$

It clearly shows that mean failure rate is reciprocal of *MTBF*.

AVAILABILITY

It is defined as the probability that a system or equipment, when used under stated conditions, without consideration for any type of maintenance shall operate satisfactorily at a given point in time does not include time spent in maintenance and other such operations. It is of two types.

(*i*) Availability (achieved): It excludes logistic time and waiting and administrative time and includes preventive and corrective maintenance down time.

(*ii*) Availability (operational): It includes ready time, logistic time and waiting or administrative down time.

The above have been expressed by the equations given below.

It is directly related with the reliability of any system or equipment.

When equipment is put-up for use after manufacture and testing, it is said to be available and when it is under breakdown it is unavailable. The term availability can therefore be defined in terms of up time and mean down time of equipment/system.

$$A = \frac{UT}{UT + MDT} \quad \text{...(1.9)}$$

Where A → availability

UT → machine/service time or up time.

MDT → machine/system down time.

Availability can be expressed in terms of meantime between failure and mean down time is given by

$$A = \frac{MTBF}{MTBF + MDT} \quad \text{...(1.10)}$$

The unavailability U is defined as the fraction of the total time test interval for which equipment is not performing to specification i.e., failed or down, therefore:

$$U = \frac{MDT}{MTBF + MDT} \quad \text{...(1.11)}$$

It is seen from above equations that availability can be improved by increasing *MTBF* i.e. how fast equipment can be repaired and put back into service.

The equation (1.9) shows availability calculations, it considers only two states of equipment/service viz working or failed, whereas there will be spectrum of intermediate states as well. For practical purposes, idle (or ready) time of machine or producı as well as the total time lost in administrative or operational delays cannot be overlooked. Thus more explicitly, the equipment's working effectiveness is expressed in terms of its operational availability (*OA*) as given below:

$$OA = \frac{OT + IT}{OT + IT + AD + RT} \quad \text{...(1.12)}$$

Where; *OT* – operating time

IT – idle time

AD – administrative or operational delays

RT – repair time.

RELIABILITY FUNCTION

The following expressions show how reliability function can be derived when the number of identical components is tested. Their success of survival and failures are estimated over well-defined time period. Here we consider a case where fixed number of identical items is tested. The number of such items is N_0. The test initiates at time $t = 0$. After a interval of time when $t > 0$, let us assume that:

$N_S(t)$ = Number of items survival at time t and

$N_f(t)$ = Number of items which have failed in time interval $(o - t)$

At time t, the reliability $R(t)$ is given by:

$$R(t) = \frac{N_S(t)}{N_o}$$

$$= \frac{N_0 - N_f(t)}{N_0}$$

$$= 1 - \frac{N_f(t)}{N_0} \qquad \text{...(1.13)}$$

Thus
$$\frac{dR(t)}{dt} = -\frac{1}{N_0}\frac{dN_f(t)}{dt}$$

As
$$dt \to 0$$

$\frac{1}{N_0}\frac{dN_f(t)}{dt}$ is the instantaneous failure density function which is expressed by $f(t)$, and

$$f(t) = -\frac{dR(t)}{dt} \qquad \text{...(1.14)}$$

Since the instantaneous hazard rate, $\lambda(t)$ is defined as the percentage of those remaining equipment that will fail in the next interval of time and is given by

$$\lambda(t) = \frac{\frac{d}{dt}N_f(t)}{N_S(t)}$$

$$= \frac{N_0}{N_0}\cdot\frac{1}{N_s(t)}\cdot\frac{dN_f(t)}{dt} = \frac{f(t)}{R(t)}$$

$$= \frac{-dR(t)}{R(t)\,dt} \qquad \text{...(1.15)}$$

From equation (1.14):

$$-dR(t) = f(t)dt$$

or
$$\int_1^{R(t)} - dR(t) = \int_0^t f(t)\,dt$$

or
$$1 - R(t) = \int_0^t f(t)\,dt$$

$$= \text{failure probability function } F(t)$$

From equation (1.15):

$$-\frac{dR(t)}{R(t)} = \lambda(t)dt$$

or
$$\int_0^{R(t)} \frac{dR(t)}{R(t)} = -\int_0^t \lambda(t)\,dt$$

or
$$\log R(t) = -\int_0^t \lambda(t)\,dt$$

or
$$R(t) = e^{-\int_0^t \lambda(t)dt} \quad ...(1.16)$$

The above equation shows that reliability and hazard/failure rates are the function of time.

From studies it is revealed that many components (electrical/electronics) have standard failure rate.

CONCEPT OF OVERALL RELIABILITY

The performance of a system can be defined in two ways, firstly the required performance, which indicates that what a system is supposed to do under all environmental conditions and secondly, the achieved performance which the system is likely to do under all environmental conditions. Overall reliability is a measure of the relationship between the complete achieved performance of the system against the corresponding required performance under all environmental conditions of space and time. Therefore, correlation of required performance and achieved performance makes the basis of the overall reliability concept and these two measures of performance may be probabilistic in nature because of their dependence on the input conditions.

In real life situations, the likely performance may have variability when different readings are taken. Under such a situation a frequency table is made to draw histograms. With larger number of variates, the histogram, which was in rectangular form, approaches some smooth curve.

The achieved performance can further be sub-divided to consider both capability of performance and variability of performance. The lack of capability may arise not only from an error in the assumed input conditions but also from any inherent design problem in the equipment itself. If by mistake incorrect data are fed to the system, the obtained reliability will also be incorrect.

The existence of variability in any practical system indicates that the value of the reliability may lie between 0 and 1 but never unity. The variations may be predictable, random or combination of both.

The following processes are essential to evaluate the reliability of a system.

- Establishment of the precise required performance of the system under all environmental conditions.
- Analysis of system to determine its inherent capability.
- Variation analysis to establish complete variability function i.e. distribution of variation.

- The process of fault mode analysis to determine availability.
- A process to obtain a particular reliability value over the appropriate time with the help of the above steps.

FACTORS INFLUENCING SYSTEM EFFECTIVENESS

The system/equipment effectiveness is indirectly related with the reliability because reliable system will help the operating personnel more effectively. If the system is effective its performance can be improved for satisfaction of the customers. An equipment may be in working condition giving sub-standard performance will not be accepted because of its poor system effectiveness. In the present day environment, the customers desire to optimize on their investments with all possible effectiveness of the system. The system/equipment will be influenced by the following factors.

(*i*) Availability of the system/equipment
(*ii*) Complexity of devices/components
(*iii*) Maintainability
(*iv*) Maintenance system practiced
(*v*) Maintenance support system
(*vi*) Failure history of the equipment/system
(*vii*) Coordination and cooperation of maintenance staff and operating personnel

Evaluation of System Effectiveness

In case of complex systems, redundant units are internally built in to help the equipment to perform their main function. Some systems can perform the same task by several different means. In this case failure of a unit can lead to decreased performance but complete system does not fail. Thus a complex system can require measure of capability that can take partial failure into account.

Effectiveness can be used to measure the performance of the system in quantitative manner. For example, quality of fuel used in IC engines can lower its performance, or poor quality service in cafeteria can lower the number of customers. The system can be said more effective with higher effectiveness index number, which can be set for different types of systems.

QUESTIONS

1. Explain the basic principle of reliability with its need in the modern industrial setup.
2. Discuss how reliability can be useful in the changing industrial system.
3. What is reliability? How it can be applied to the existing system?
4. Define availability with its importance to maintenance organization.
5. How reliability functions can be useful for reliability assessment of system?
6. Discuss the concept of overall reliability.
7. Enlist the factors responsible for system effectiveness.

2

Statistics for Reliability

INTRODUCTION

The use of statistics in reliability is very important since it is associated with probability and other parameters, which are variables. The reliability assessment can only be made from field data or some laboratory experiments. The components/systems which are either manufactured or fabricated may not have the same life. The performance of each component depends on many factors such as manufacturing defects, strength of the material used and other environmental conditions such as temperature, humidity etc. Since, reliability depends on the factor which are variable, a knowledge of statistical nature of variables is necessary. The data collected for the study and analysis will play a vital role since the final results obtained will depend on them. The number of failures and the repair times of the equipment/system will be in the form of number or variable, therefore, knowledge about it, is essential. The facts expressed in quantitative form can be termed as data, which again is a variable. In this chapter we will discuss some aspects of statistics related to reliability analysis.

RANDOM VARIABLES

All the events are initially associated with real numbers and later on with the probabilities. The outcome of an event may be a integer number such as 0,1,2,3 and so on but event corresponding to each value is evident. The function, which assigns numbers to events, is called as random variable. This is neither random nor a variable in real sense. In the real world random variable is a ordinary variable which is selected randomly. In other words, once a random variable is well defined, we may speak of any value in the range of the random variable as if it were actually the event. It makes sense, thereby, to speak of the probability that a random variable, X, equals some particular number. It is important to note that, values of random variables have probabilities associated with them only because the values correspond to events, which deserve the probability directly.

The random variables are of two types, the first being discrete, where the variables are integers or more precisely, a countable subset of real numbers. Second one, being continuous, where the range consists of all values over an interval of the real numbers. By random variable we mean that each number of the sampled population should have an equal chance of being selected.

For example to test the strength of material, standard test specimen that are machined from randomly selected bars are then tested. Their dimensions may vary though made by the trained technicians and therefore the test results, since the testing is done on a random specimen, the experiment is random experiment and outcome of the result is a random variable. These results can be expressed in many ways.

FAILURE DISTRIBUTION

During the reliability analysis of an equipment/system it is essential to know its failure pattern. In some cases it is not possible to get actual information regarding failures of components and sub assemblies. In such cases it is desired to know the nature of the failure or in other words the failure distribution. The failure pattern can be obtained from field data, which are collected from the history cards of the equipment/systems. In some cases it is desired to identify the failure indicating parameters before actual failure of the equipment/system takes place, as the past information on the subject matter is not available. The methods employed for ascertaining such values are termed as hazard/failure functions and are of immense help for the calculation of reliability. This will also help to determine failure density functions and failure distributions. The cause of failures can be anything predictable or unpredictable. These can be represented graphically in the form of histogram or a frequency bar diagram.

CONCEPT OF PROBABILITY

Reliability is directly associated with probability, therefore its basic concept needs to be discussed at this point before discussing the various distribution involved in the study. Uncertainty is attached with human life. Decision-making in such areas is facilitated through formal and precise expressions for the uncertainties involved. Probability theory provides us with the ways and means to attain the formal and precise expressions for uncertainties involved under different situations.

Probability measures the uncertainty about the occurrence of a particular event or a set of events and is expressed numerically between zero and one. This can be estimated by any of the three methods given below:

(*i*) Objective approach—which could be classical or empirical. These are based on natural laws or on the experiments performed actually and don't depend on individual opinions.

(*ii*) Subjective approach—where probability measures the degree of confidence, which ultimately depends upon the occurrence and authenticity of the information used and plays important role in making decisions.

(*iii*) Modern approach—which combines both the approaches cited above and are based on the theory of sets. This can be expressed through Venn diagram and tree diagrams. The following terms are associated with probability.

Experiments

These are different than those carried out in physics or chemistry laboratories. Here any action taken to specify the events is called experiment, which has three common parameters. The first being the outcomes of the experiments which may be two or more in number and secondly it is possible to specify outcomes in advance. The last item is that the outcomes are with certain uncertainty.

Sample Space

The set of all possible outcomes of an experiment is defined as the sample space. Each outcome is visualized as a sample point in the sample space. It is virtually the field from where experiments are carried out, with certain outcomes.

Event

In probability theory, it constitutes one or more possible outcomes of an experiment. For example, tossing of coin i.e. head or tail. It can be defined as a sub-set of a sample space. Here event is used to refer to a single outcome or combination of outcomes.

The axioms of the probability are the following:

(*i*) The probability of an even A, written as $P(a)$, must be a number between zero and one, both values inclusive.

Thus $$0 \le P(A) \le 1$$

(*ii*) The probability of occurrence of one or the other all possible events is equal to one. As S denotes the sample space therefore, $P(s) = 1$.

(*iii*) If two events are such that occurrence of one rules out occurrence of another then the probability that either one or the other will occur is equal to the sum of their individual probabilities.

With above given axioms probability is defined as a function, which assigns probability value P to each sample point of an experiment abiding by the above axioms.

Probability Distributions

Any rule, which assigns probabilities to each of the possible values of random variable, is a probability distribution. For discrete random variables the rule is to indicate the probability of each value separately. The function $p(x)$ is defined as:

$p(x) = P(X = x)$, and is called probability distribution function. It can also be expressed as cumulative distribution function $F(x)$ defined as:

$$F(x) = P(X \le x)$$

It can also be expressed; in form of complementary cumulative function $G(x)$ which is expressed as

$$G(x) = P(X > x)$$

If any one of the above is known the others can be obtained.

For continuous random variable the situation is different because the range of possible values is infinite and is not consistent with axioms of the probability theory. Therefore particular value to these random variable are difficult to assign.

Following are the commonly used distributions for failure/repair analysis of engineering systems. These distributions enable us to estimate parameters based on which statistical analysis of the equipment/system performance is computed and probabilistic/expected values found, concerning, mean time to failure, mean time to repair, availability and maintainability functions and reliability indices. Perhaps exponential and Weibull distribution are the most widely used distributions for life testing applications.

NORMAL DISTRIBUTION

It is the most important continuous probability theoretical distribution and has been useful in countless applications involving every conceivable discipline. The theoretical distribution is an ideal representation of the real life behaviour of many mechanical components and systems. When the real behavioural patterns are represented in some mathematical forms, some common features can be extracted and these features display the similarities among diverse components and systems. A number of properties of this distribution make it easy and useful for mathematical applications. The probability density function (*p d f*) for this distribution is defined by the equation given below:

$$f(x) = \frac{1}{\sigma\sqrt{2\pi}} e^{-[(x-\mu)^2/2\sigma^2]} \qquad \text{...(2.1)}$$

$$-\infty \le x \le \infty$$

Where μ is unrestricted parameter and σ is positive parameter and *x* is the real number denoting random variable.

Here variable μ and σ are the parameters to specify mean and standard deviation of a random variable. Any linear transformation of a randomly distributed random variable is also normally distributed.

The cumulative density function is expressed as:

$$F(X) = \int_{-\infty}^{x} \frac{1}{\sigma\sqrt{2\pi}} e^{-\frac{1}{2}\left(\frac{y-\mu}{\sigma}\right)^2} dy \qquad \text{...(2.2)}$$

It is apparent from above that *f* is a positive function, $e^{-\frac{1}{2}\left(\frac{x-\mu}{\sigma}\right)}$ being positive for any real number *x*. It can be shown that $\int_{-\infty}^{+\infty} f(x)dx = 1$ so that *f*(*x*) is a valid p.d.f. The plot of *f*(*x*) and *x* gives a well spread curve symmetrical about the population mean μ and area under it is unity. For smaller values of σ the curves are peaked and for larger values of *x* the curve becomes broader. These are shown in Fig. 2.1.

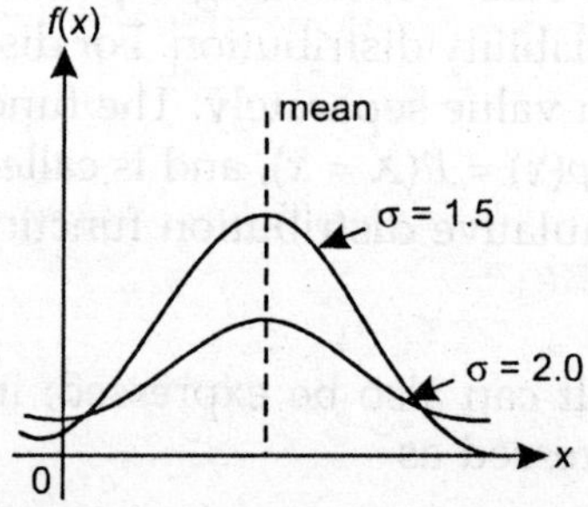

Fig. 2.1: Normal Distribution Curve

Every normal distribution with, whatever values of parameters can be represented in terms of the standard normal distribution which has a mean of zero and variance of 1. The linear transformation required to convert a normally distributed random variable *X* with mean μ and variance σ^2 to the standard normal random variable *Z* is

$$Z = \frac{X-\mu}{\sigma}$$

The density function of the standard normal random variable is just

$$f(x) = \frac{1}{\sqrt{2\pi}} e^{-\left(\frac{x^2}{2}\right)} \quad -\infty \le x \le \infty \qquad ...(2.3)$$

The normal distribution is reproductive that is the sum of the two or more normally distributed random variables is itself normally distributed. The mean of the sum, is always the sum of means. The variance of the sum is the sum of the variances provided that the random variables are independent. Even if they are not, the variance of sum can be expressed in terms of the variances and co-variances of the constituents.

The famous central limit theorem states, that the sum of the large number of independent arbitrary distributed random variable will be (approximately) normally distributed (under certain condition).

This distribution is not frequently used in reliability studies however, the failure density in wear out region in case of power system components is often represented by the normal distribution.

THE LOG NORMAL DISTRIBUTION

It is the distribution of a random variable whose natural logarithm follows a normal distribution. The log normal density function is expressed as:

$$f(x) = \frac{1}{\sigma x\sqrt{2\pi}} e^{[(\ln x - \mu)^2/2\sigma^2]} \qquad ...(2.4)$$

Where $x > 0$

The parameter μ and σ may given by:

$$\mu = E(\ln X)$$

$$\sigma = V(\ln X)$$

But the mean and variance of X are, respectively

$$E(X) = e^{\mu + (1/2\sigma^2)}$$

$$V(X) = e^{2\mu + \sigma^2}(e^{\sigma^2} - 1)$$

In practice, the best way to deal with a log normally distributed random variable X is to transform it by taking its natural logarithm. That is, let $Y = \ln X$, of course, Y would be normally distributed and is therefore easily handled.

The log normal distribution arises from the product of many independent non-negative random variables, as against in normal it is the sum of the independent random variables.

Under the following conditions the above distribution is useful.

(*i*) The distribution is not symmetrical about the mean and

(*ii*) The variables have only positive values.

The fact why this distribution is used under above conditions is that while the random variable considered is by itself not normally distributed, its logarithm is normally distributed.

The examples of these are compressive strength of concrete, fatigue life of mechanical components, and wear life of bearings, since random variables follow log normal distribution.

THE POISSON DISTRIBUTION

This distribution is more useful where number of trials or experiments is high and the probability of occurrence of an event is small. The Poisson distribution is the special case of Bernoulli's distribution. All that is required for the approximation is to give the distribution the same expectation. The general expression for this can be given as follows:

$$f(x)= \frac{\lambda^x e^{-\lambda}}{x} \quad \text{Where} \quad x = 0, 1, 2, 3 \text{-----------} \infty \qquad ...(2.5)$$

X has a Poisson distribution with parameter λ, which must be positive. The important property of this distribution is that the expectation and variance are equal to each other.

$$E(x) = V(x) = \lambda$$

This distribution is productive, which means that the sum of Poisson distributed random variables will be another Poisson distribution.

Thus the Poisson distribution represents the probability of an event occurring, a specified number of times in a given time interval or space when the rate of occurrence is fixed.

EXPONENTIAL DISTRIBUTION

For a continuous random variable the probability density function of exponential distribution can be expressed as:

$$f(x) = \lambda\, e^{-\lambda x} \qquad x \geq 0, \lambda \geq 0 \qquad ...(2.6)$$

The cumulative distribution function has the following expression

$$F(x) = 1 - e^{-\lambda x} \qquad ...(2.7)$$

The complementary cumulative distribution function is expressed as:

$$G(x) = e^{-\lambda x} \qquad ...(2.8)$$

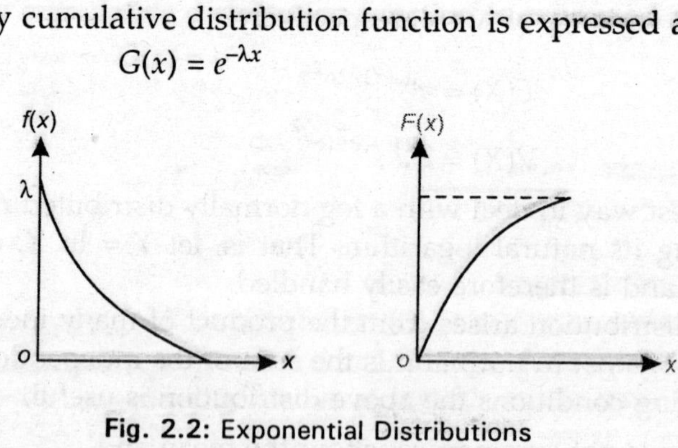

Fig. 2.2: Exponential Distributions

The expectation of negative exponential distributed random variable is $E(x) = \frac{1}{\lambda}$ which is reciprocal of the parameter and the variance is $V(x) = \frac{1}{\lambda^2}$.

The time between breakdown of machines and life of an electric bulb etc are the some examples where the exponential distribution has been found useful. We may be interested in the random variable of the lapse of time before the first occurrence of event. Thus, for a machine, we note that the first failure may occur after a month or 1.5 months. The number of failures within a specified period is discrete and follows Poisson distribution, whereas the variable, time of first failure is continuous and will follow exponential distribution since it represents uncertainty.

THE WEIBULL DISTRIBUTION

This is an important distribution used in the analysis of reliability of a equipment/system. By properly selecting the parameters the curve obtained can represent a variety of experimental/field results. It also represents a good approximation of normal distribution and also exponential distribution.

The Weibull distribution is of two types i.e., two parameters and three parameters and their corresponding density functions are given as below:

$$f(x) = \frac{m}{\theta}\left(\frac{x}{\theta}\right)^{m-1} e^{\left[-\left(\frac{x}{\vartheta}\right)^{m}\right]} \quad x \geq x_0 \qquad ...(2.9)$$

$$f(x) = \frac{m}{\theta - x_0}\left(\frac{x - x_0}{\theta - x_0}\right)^{m-1} e^{\left[-\left(\frac{x - x_0}{\theta - x_0}\right)^{m}\right]} \quad x \geq x_0 \qquad ...(2.10)$$

Where θ = characteristic or scale value ($\theta \geq 0$)

m = shape parameter

x_0 = guaranteed value of x ($x_0 \geq 0$)

If x_o is equal to zero, the three parameter density function reduces to two parameter function.

The above Fig. 2.3 shows the density function for different values of m. It is bounded by x_o the lower tail and therefore x_o is called the guaranteed variate. The shape parameter controls the skewness of the distribution. When the values of x falls within the range $3.3 \leq m \leq 3.5$ the symmetry is displayed approximately with good fitness of normal distribution. When m equals to one it gives exponential distribution. The cumulative distribution function can be represented by the expression given below:

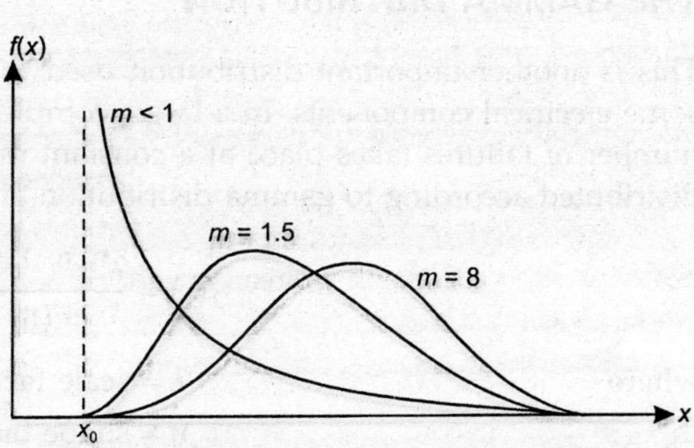

Fig. 2.3: Weibull Distributions

$$F(x) = e^{\left[-\frac{(x - x_0)^{m}}{(\theta - x_0)}\right]} \qquad ...(2.11)$$

But putting $x_o = 0$ the cumulative distribution t is obtained for two parameters and expression will be

$$F(x) = 1 - e^{\left[-\left(\frac{x}{\theta}\right)^m\right]} \qquad ...(2.12)$$

The mean of the Weibull distributed random variable is

$$\mu_x = x_0 + (\theta - x_0)\,\Gamma\left(1 + \frac{1}{m}\right) \qquad ...(2.13)$$

and standard deviation is expressed as:

$$\hat{\sigma}_x = (\theta - x_0)\left[\Gamma\left(1 + \frac{2}{m}\right) - \Gamma^2\left(1 + \frac{1}{m}\right)\right]^{1/2} \qquad ...(2.14)$$

Where Γ is the gamma function?

The median of the distribution is expressed as:

$$X_{med} = x_o + (\theta - x_0)\,(\ln 2)^{1/m} \qquad ...(2.15)$$

and mode is represented by,

$$X_{mode} = x_0 + (\theta - x_0)\left(\frac{m-1}{m}\right)^{1/m} \qquad ...(2.16)$$

This distribution has been useful for describing life time and waiting time in reliability applications. If in a system there are many components, which have lifetime distribution of their own and if the equipment fails as any one of the component fails, then the lifetime of the equipment/system is minimum of the life times of its components. Under these conditions, there is theoretical justification of expecting a Weibull distribution to provide close approximation to the lifetime distribution of the equipment/system.

THE GAMMA DISTRIBUTION

This is another important distribution used in reliability analysis particularly for failure of some electrical components. In a large, complex engineering system, if the shape factor of a number of failures takes place at a constant rate, then the failure of a system as a whole is distributed according to gamma distribution and can be expressed as:

$$f(x) = \frac{\lambda^n x^{n-1} e^{-\lambda x}}{\Gamma(\eta)} \text{ for } x, \lambda, \eta \geq 0 \qquad ...(2.17)$$

where λ = scale factor

η = shape factor and

$$\text{Gamma function } \Gamma(x) = \int_0^{\infty} x^{n-1} e^{-x} dx$$

Figure 2.4 shows the gamma density function for different value of λ and η = 3 when η = 1 the density function reduces to $f(x) = \lambda e^{-\lambda x}$ which is the exponential distribution.

For given value of the values of Γ(η) can be found from the table. If η is an integer, Γ(η) = (η − 1)!. In this case gamma distribution reduces to the Erlang. Therefore gamma distribution is a special case of Erlang distribution. Under special conditions this distribution also represent Chi-square density function.

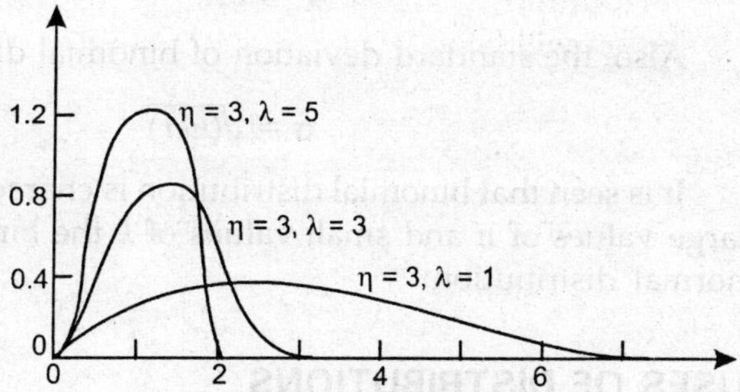

Fig. 2.4: Gamma Distribution

The expectation and variance of gamma distributed random variable *X* is expressed as:

$$E(X) = \frac{\eta}{\lambda}$$

$$V(X) = \frac{\eta}{\lambda 2}$$

THE BINOMIAL DISTRIBUTION

This distribution describes the situation where either of the two results i.e. success or failure are possible outcome of an experiment/trial. If *K* is the probability of success in each trail the failure probability is $r = 1 - k$. Such repeated independent trials are called Bernoulli trials since there are only two possible outcomes for each trial/experiment. Their probability remains same throughout the trials/experiments. Then, the probability of success of *x* number of experiments out of *n* trials follows the binomial distribution law given by the following expression.

$$p(X = x) = C_x^n K^x r^{n-x} \qquad \text{...(2.18)}$$

Where $$C_x^n = \frac{n!}{x(n-x)!} \qquad \text{...(2.19)}$$

From equation 3.18 the probability of no success is

$$p(X = 0) C_x^n K^0 r^n = r^n$$

and probability of at least one success is

$$p(x \geq 1) \;= p(x = 1 + p(x = 2) + p(x = 3) + \text{-------}$$

since $$p(x = 0) + p(x = 1) + p(x = 2) + \text{-----} p(x = n) = \Sigma_0^n C_n^x K^x r^{n-x} = 1$$

$$p(x \geq 1) = 1 - p(x = 0) = 1 - r^n$$

The ratio x/n is called as proportion and mean of the binomial distribution can be shown to be

$$\overline{X} = nK$$

Also, the standard deviation of binomial distribution is found to be

$$\hat{\sigma} = \sqrt{(nkr)}$$

It is seen that binomial distribution is characterized by two parameter namely n and k. For large values of n and small values of k the binomial distribution can be approximated to a normal distribution.

USES OF DISTRIBUTIONS

Data collected from either experiment or field trials have to check for their approximate distribution since data are random in nature. Sometimes it is difficult to collect sufficient data for accurate analysis, knowing the pattern or distribution the same can be generated to follow the family appropriately. Through Bernoulli trails the success or failure of an event can be examined and this can fit under binomial distribution. In case of the large number of independent random variables, the central limit theorem can suggest the normal distribution. On the other occasions; the choice of distribution is influenced by the need for particular mathematical properties. The negative exponential distribution in Markov process models are selected for reasons that have little to do with observed data.

To really know the nature of data a histogram is a first step in identifying an appropriate distribution. This will indicate the shape and later it has to be fixed with nearest P.D.F. once at least a distribution type is fixed, the next problem is to set value for the parameter that fix the distribution within the family. In certain cases it is easy, for example in case of Poisson distribution. Parameter λ is estimated by sample mean μ and σ^2 in case of normal distribution by the sample means and sample variance respectively. Gamma and Weibull distributions require additional statistical work before proper estimation of formulae. The next step is to validate the model by checking the goodness of fit. For this, statistical tests such as Chi-square and Kolmogorove-Smirnov can be used for goodness of fit. The null-hypothesis assures that the candidate distribution is correct however. Care should be taken for the quantity of data, because the tests carried out, may not yield correct results, if database is small. Only the large quantity of data can give satisfactory results with the aid of hypothesis.

QUESTIONS

1. Highlight the importance of statistics in view of reliability evaluation.
2. Define random variable and state how they are generated for experimentation.
3. What is the significance of failure distributions in the context of reliability?
4. How probability is associated with reliability? Discuss briefly.
5. Discuss the distributions which can be used for reliability analysis.
6. Why normal distribution has wide applicability in mathematical applications?
7. Explain why Weibull distribution is more appropriate in the case of reliability evaluation.

3

Reliability of Systems

INTRODUCTION

In Chapter 1 the relationship of reliability with failure probability, failure rate, mean time between failures etc have been discussed. Most of the engineering systems are made up of several simpler components/elements. Their failure in due course of time is inevitable for various reasons such as wear, stress concentration or even breakage etc. As the systems are made up of components and units, the performance of these has to be integrated and coordinated so as to make the system work satisfactorily for a specified time period. The greater the number of elements involved larger is the system and its complexity, which will make the reliability analysis difficult. The reliability of the complete system will depend upon the reliability of the each element/component involved. This chapter deals with calculation of reliability for all complex systems presently used in industrial organizations. The models used for analysis of reliability are briefly discussed as under.

SERIES RELIABILITY MODEL

When the success of a system depends on the success of all its elements/components, this gives rise to a series reliability model. Consider the reliability block diagram of Fig. 3.1 in which n components with reliabilities as R_1, R_2, R_3 ---------- R_n is connected in series. In some cases the physical connection of the elements/equipment may not be in series but, if failure of one component brings the system to halt, is also a case of series system. The practical example of this is the connection of batteries in series for more power. However, the cylinders placed in internal combustion engines may have series configuration but the failure of one cylinder will not make the engine completely out of order though its performance will be reduced.

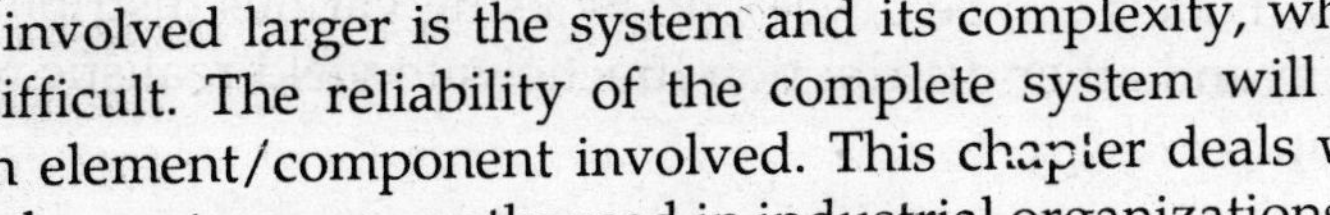

Fig. 3.1: Series Reliability Model

The system reliability R_{sys} is therefore the product of the individual reliabilities, i.e.

$$R_{sys} = R_1 \cdot R_2 \text{ -------------- } \cdot R_n \qquad ...(3.1)$$

If we assume further that each component can be described by a constant failure rate λ and if λ_i is the failure rate of ith component, then reliability of ith component, R_i can be expressed by

$$R_i = e^{-\lambda_i t} \quad ...(3.2)$$

Thus $$R_{sys} = e^{-\lambda_1 t} \cdot e^{-\lambda_2 t} ------ e^{-\lambda_n t} \quad ...(3.3)$$

So that if γ_{sys} is the total system failure rate then:

$$R_{sys} = e^{-\lambda_{sys} t} = e^{-(\lambda_1 + \lambda_2 + --- + \lambda_n)t} \quad ...(3.4)$$

and $$\lambda_{sys} = \lambda_1 + \lambda_2 + ------ + \lambda_n$$

This means that the overall failure rate for a series system is the sum of the individual element or component failure rates. It is seen from above equations that if the number of components is minimum, the system reliability will be maximum.

The reliability can also be expressed in terms of unreliability and if $F_1, F_2 ... F_n$ are the unreliability's of the elements 1 to n and the system unreliability be F_{sys}. Then system reliability will be

$$R_s = 1 - F_{sys} \quad ...(3.5)$$

or $$1 - F_{sys} = (1 - F_1)(1 - F_2) ------------ (1 - F_n)$$

The unreliability of a series system with small values of unreliability's can be given by:

$$F_{sys} = [F_1, F_2 ------ F_n] \quad ...(3.6)$$

Series reliability models can be developed by using various distributions such as normal, exponential, uniform and other distribution functions to yield realistic results for mechanical systems/equipment.

Series Reliability—Components in Triplicate

Consider a system with indentical components connected as shown in Fig. 3.1(a).

$$R = [(1 - R_1)(1 - R_2)(1 - R_3)];\ 1 - (1 - R_1)^3$$

where $R_1 = R_2 = R_3$

$$= 1 - (1 - 3R_1 - 3R_1^2 - R_1^3)$$

$$= 3R_1 - 3R_1^2 - R_1^3$$

$$= 3e^{-\lambda_1} - 3e^{-2\lambda_1} + e^{-3\lambda_1}$$

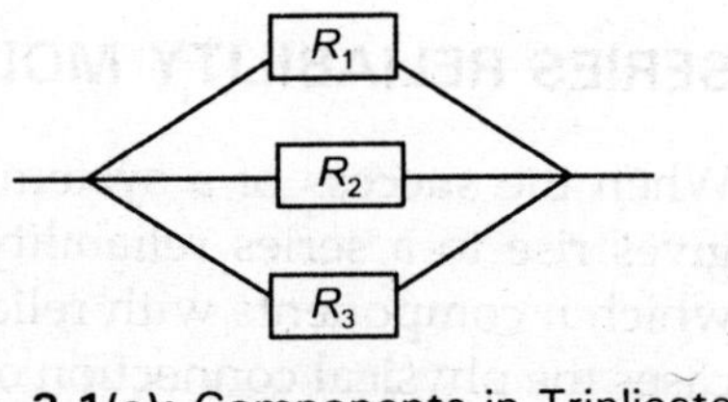

Fig. 3.1(a): Components in Triplicate

PARALLEL RELIABILITY MODEL

When a system of n components can function even if only one of the components is good, this gives rise to a parallel reliability model (fully redundant) [the example being multi cylinder engine]. Consider the block diagram vehicle of Fig. 3.2 in which elements (components having reliability as $R_1, R_2 --- R_n$ are connected in parallel). The product law of unreliability's derives the reliability of this system more conveniently.

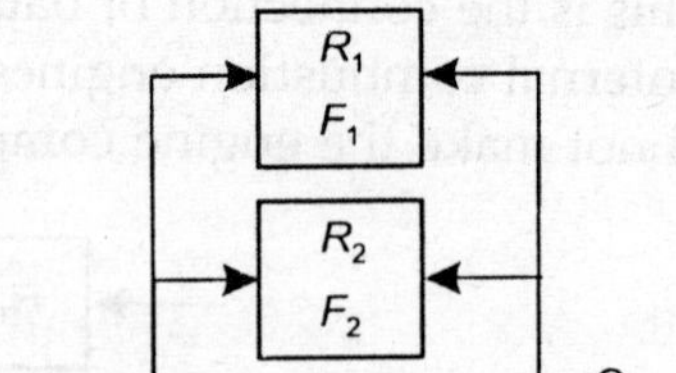

Fig. 3.2: Parallel Reliability Model

$$R_{sys} = 1 - \prod_{i=1}^{n}(1 - R_i) = 1 - \prod_{i=1}^{n}(1 - e^{-\lambda_i t}) \quad ...(3.7)$$

The reliability can also be expressed as given below:

Consider that there are n individual elements/components connected in parallel with individual unreliability's F_1, F_2 - - - F_n respectively. All of the elements/components are expected to be working normally i.e. to be active and each one is capable of meeting the functional requirements placed on the overall system. The overall system will only fail when all of its elements/components fail. If one element or component survives the system survives. The overall system unreliability F_{sys} is therefore the product of the individual element/component unreliability i.e.

$$F_{sys} = F_1 F_2 .. F_n \quad ...(3.8)$$

Comparing the equations 3.1 and 3.8 it is seen that for series system, the system reliability is the product of element/component reliabilities, whereas for parallel systems, unreliability is the product of element/component—unreliability's. If the individual elements/components are identical so that $F_1 = F_2 - - F_n = F$. This gives

$$F_{sys} = F^n$$

Thus if, $F = 0.1$ and there are four channels in parallel, we have $F_{sys} = 10^{-4}$. It is seen that by increasing the number of elements/components in parallel system, increases the overall system reliability.

For example, consider a system with four elements whose failure-free performance probabilities are:

$$p(x_1) = 0.90,\ p(x_2) = 0.91,\ p(x_3) = 0.94 \quad \text{and} \quad p(x_4) = 0.92$$

Therefore, the system reliability is given by

$$R(s) = 1 - [1 - 0.90] \cdot [1 - 0.91] . [1 - 0.94] . [1 - 0.92]$$
$$= 1[0.1 \times 0.09 \times 0.06 \times 0.08] = 0.9999568$$

In case of r out of n system can be handled using the Binomial distribution. The general equation of redundant elements assumes that the elements are stochastically independent. Success of exactly r out of n identical independent components is given by,

$$R_S(r/n) = {}^nC_r R^r (1 - R)^{n-r}$$
$$= {}^nC_r e^{-r\lambda t}(1 - e^{-\lambda t})^{n-r} \quad ...(3.9)$$

and the success of at least r out of n elements is given by

$$R_S(r \text{ and more}/n) = \sum_{k=r}^{n} {}^nC_k R^k (1 - R)^{n-k}$$

$$= \sum_{k=r}^{n} {}^nC_k e^{-k\lambda t}(1 - e^{-\lambda t})^{n-k} \quad ...(3.10)$$

SERIES PARALLEL RELIABILITY MODEL

The parallel element model shown in Fig. 3.2 is also sometimes referred as active redundant system. Here active redundancy means that all elements are active and not kept in reserve. The case where elements are kept under reserve will be dealt with separately. The Figure 3.3 shows elements in series and in parallel also.

It is a very common model being used in any engineering equipment where n identical sub-systems are placed in parallel and each system consists of m elements/components in series. If R_{ji} is the reliability of the ith element in j^{th} subsystem then reliability of the jth subsystem is:

$$R_j = R_{j1}\, R_{j2} \text{ -------- } R_{ji} \text{ -------- } R_{jm} = \prod_{i=1}^{i=m} R_{ji} \qquad \text{...(3.11)}$$

The corresponding unreliability of the jth subsystem is:

$$F_j = 1 - \prod_{i=1}^{i=m} R_{ji} \qquad \text{...(3.12)}$$

The overall unreliability of the system is expressed by the following equation:

$$F_{overall} = \prod_{j=1}^{i=n} \left[1 - \prod_{i=1}^{i=m} R_{ji} \right] \qquad \text{...(3.13)}$$

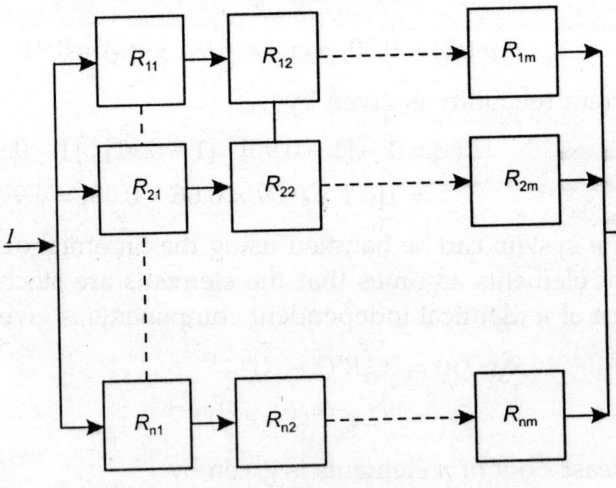

Fig. 3.3: Series Parallel Reliability Model

In the usual practice of obtaining the reliability expression for a series parallel system, the sub system with parallel elements are individually reduced to single-element units by applying the rule of probability. Then the system is modified to one consisting of equivalent elements, all connected in series and the final solution can be obtained using series reliability model.

DECOMPOSED RELIABILITY MODEL

In practice a complex system configuration is such that neither a pure series nor a pure parallel reliability model is appropriate. In this case the reliability structure of the complex system is decomposed into simpler sub-structure through successive application of conditional probability theorem. The technique starts with the selection of a key component/element which appears to bind together the overall reliability structure, and the reliability may be expressed in terms of the key component as:

$$R_s = P\text{ (system success if } B \text{ is good)} \cdot P(Bx) + P\text{(system success if } B \text{ is bad)} \cdot P(By) \quad ...(3.14)$$

Where $P(Bx)$ and $P(By)$ are respective success and failure/probabilities of the key component B. In equation (3.14) one difficult problem is decomposed into two easier ones. In a complex problem the decomposition process is repeated on the sub-structures formed after first decomposition and the system reliability is obtained sequentially. If a poor choice is made for the key component, the method still works well; but the derivation of reliability becomes cumbersome.

REDUNDANT RELIABILITY MODEL

In many complex equipment/systems it is difficult to improve the reliability, by just selecting the good quality components only. In many cases it is desired to have a redundant and back up elements and subsystem to improve or enhance the reliability of the total system. Sometime safety aspects and hazardous conditions demand the use of redundant components or sub-systems. Automobiles carrying spare wheel are the example of such systems.

There are three types of redundancies that are applied to the system. Which include:

(*i*) Active redundancy, where all the elements/components used are active. This also called as hot redundancy.

(*ii*) Standby redundancy, in this case duplicate or spare elements/systems are used in case of failure of main or the active components. Places where working on equipment/systems are risky the concept of standby is always employed. This may also be termed as cold redundancy.

(*iii*) Light duty redundancies—this system is employed where the variations in load are experienced. Under such conditions spare or duplicate equipment/systems are kept under operation during light load conditions but can also be used when the active system fails. This is also called as worm redundancy. In the power generation area it is the most common practice to meet out the fluctuating loads.

Active Redundancy

Under this system all the redundant element components function along with the main ones and the system will only fail after all its element/components have failed. Figure 3.4(a) and (b) show two parallel systems where each element has been duplicated and connected with the main system.

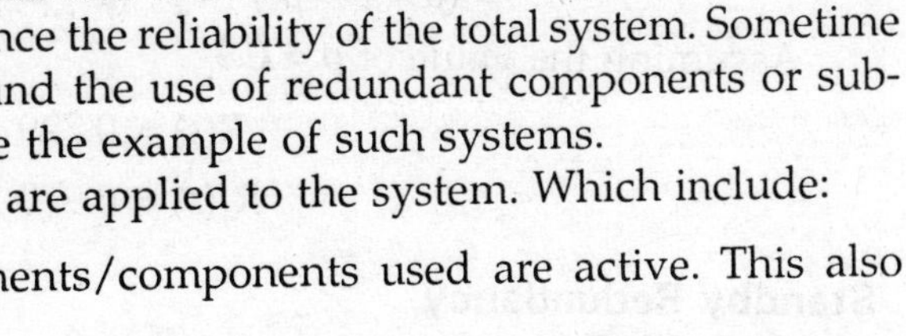

Fig. 3.4(a)

Let $p(A)$, $p(B)$, be the probabilities of failure free performance of the components and $R(s)$ is the probability of failure free performance of the system then from Fig. 3.4(a).

$$R(s) = p(A \text{ and } B \text{ and } C) \text{ or } p(A_1 \text{ and } B_1 \text{ and } C_1)$$
$$= p(A) \cdot p(B) \cdot p(C) \text{ or } p(A_1)\, p(B_1) \cdot P(C_1)$$
$$= p_a p_b p_c + p_{a1} p_{b1} p_{c1} - p_a p_b p_c p_{a1} p_{b1} p_{c1}$$
$$= p^3 + p^3 - p^6 = p^3(2 - p^3) \text{ for identical components} \quad ...(3.15)$$

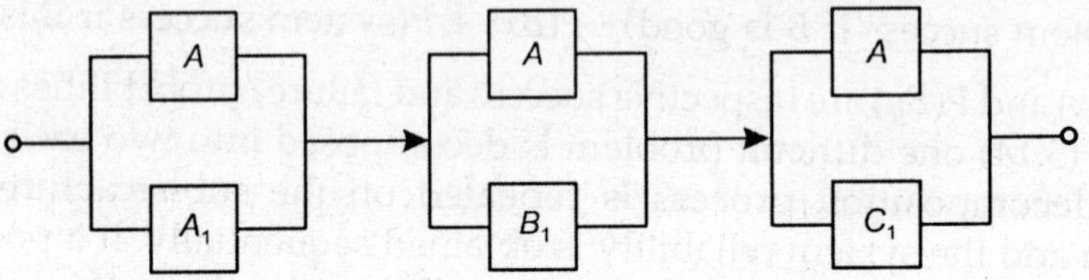

Fig. 3.4(b)

Similarly for Fig. 3.4(b)

$$R(s) = p(A \text{ or } A_1) \text{ and } p(B \text{ or } B_1) \text{ and } p(C \text{ or } C_1) \text{ and } p(D \text{ or } D_1)$$

Again considering identical components

$$= (p + p - p^2)^3 = p^3\,(2 - p)^3 \quad ...(3.16)$$

Assuming the value of $p = 0.9$

$R(s) = 0.729$ for Fig. 3.4(a)

$R_1(s) = 0.970$ for Fig. 3.4(b)

Standby Redundancy

Two components A and B are connected in parallel Fig. 3.4(c) and both components are active. The Reliability of failure free performance is expressed as

$$R(s) = p(A) \quad \text{or} \quad p(B) \quad ...(3.17)$$

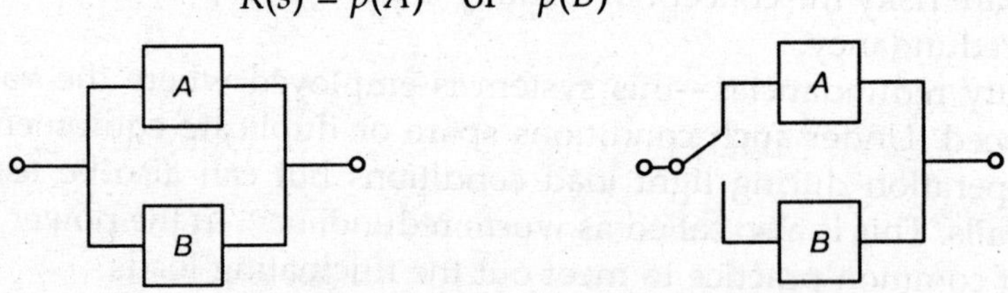

(c) (d)

Fig. 3.4(c and d)

In Fig. 3.4(d) the component A is only active and the component B is kept under reserve and activated by a decision switch only when A fails. The switch used is considered reliable and operates instantaneously. The system can work under the following conditions:

(*i*) Component A does not fail within the specified time t or

(*ii*) Component A fails at time $(t_1 \leq t)$ and the component B operates without failure from time t_1 to time t.

The probability that component A does not fail with the given time t is expressed as:

$$R_1(t) = 1 - F_a(t) = \int_a^{\infty} f_a(\tau)\, d(\tau) \qquad ...(3.18)$$

Where $f_a(\tau)$ is the failure density function for A. The probability that the component A fails in the interval $(t_1 \pm \Delta t_1/2)$ is $f_a(t_1)\Delta t_1$; the probability that the component B functions failure free from time t_1 to time t is given by the expression:

$$\int_{t-t_1}^{\infty} f_{(b)}(\tau)\, d\tau \qquad ...(3.19)$$

Where $f_b(\tau)$ is the failure density function of the component B. Therefore the probability for both above conditions can be given as:

$$f_a(t_1)dt_1 \int_{t-t_1}^{\infty} f_{(b)}(\tau)\, d\tau$$

Since t_1 and vary from 0 to t the probability for case (*ii*) then becomes

$$R_i(t) = \int_0^1 f_a(t_1) \left[\int_{t-t_1}^{\infty} f_b(\tau) d\tau \right] dt_1 \qquad ...(3.20)$$

Since case (*i*) and (*ii*) are mutually exclusive the total probability is

$$R(t) = R_1(t) + R_2(t)$$

$$= \int_t^{\infty} f_a(\tau) d\tau + \int_0^t f_a(t_1) \left[\int_{t-t}^{\infty} f_b(\tau) d\tau \, dt_1 \right] \qquad ...(3.21)$$

In the similar way the probability calculations can be done when more than one standby components are considered.

MARKOV RELIABILITY MODEL

The reliability problem normally deals with systems that are discrete in nature and continuous in time in order to formulate a Markov model first defines all the mutually exclusive state of the system. For example, in a system of a single non-repairable element, there are two possible states, UP state when the element is operable and DOWN state when the element has failed. The states of the system at $t = 0$ are called initial states, and those representative of a final or equilibrium state are called the final state. The state space diagrams for this system are shown in Fig. 3.5.

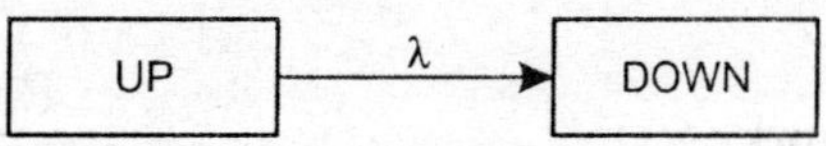

Fig. 3.5: State Space Diagram of Non-repairable Unit

The equations, which describe the probabilistic transitions from the initial to the final states, are known as Markov state equations. The transition probabilities follow two assumptions:

(*i*) The probability of transition in time Δt from one state to another is given by $\lambda(t)\ \Delta t$, where $\lambda(t)$ is the hazard associated with the two states. If all the $\lambda_i(t)$'S are constant, the model is known as homogenous. If any hazards are time functions, the model becomes non-homogenous.

(*ii*) The probabilities of two or more transitions in time Δt are considered negligible.

Assume:

$P_o\ (t)$ = Probability that the component is being in UP state at time (t).

$P_1\ (t)$ = Probability that the component is being in DOWN state; and

λ = Failure/hazard rate.

Then the probability that the component is in the UP state at time $t + \Delta t$, $P_0(t + \Delta t)$ is given by the product of the probability of the system in being UP state at time t, $P_0\ (t)$ and probability of no failure in time Δt, $1 - \lambda\Delta t$, plus the probability of being in DOWN state at time t, $P_1\ (t)$, times the probability of transition to the UP state (with repair) in time Δt. Thus,

$$P_0(t + \Delta t) = (1 - \lambda\Delta t) \cdot P_0(t) + 1.0P_1(t) \quad ...(3.22)$$

And probability of being in DOWN state at $t + \Delta t$ is given by

$$P_1(t + \Delta t) = 1 \cdot P_1(t) + \lambda\Delta t \cdot P_0(t) \quad ...(3.23)$$

The equation (3.22) and (3.23) can be summarized by writing the transitional probability matrix for the above system as given in Table 3.1.

Table 3.1: State Transition Matrixes

	0	1
0	$1 - \lambda\Delta t$	$\lambda\Delta t$
1	0	1

It may be noted that in a transition matrix its rows sum must be unity. Rearranging equation (3.22) and (3.23) gives:

$$\frac{P_0(t + \Delta t) - P_0(t)}{\Delta t} = -\lambda P_0 t$$

$$\frac{P_1(t + \Delta t) - P_1(t)}{\Delta t} = \lambda\ P_1 t$$

and

$$\frac{P_0(t + \Delta t) - P_0(t)}{\Delta t} \downarrow_{\Delta t \to 0} = P_0^l(t)$$

$$\frac{P_1(t + \Delta t) - P_1(t)}{\Delta t} \downarrow_{\Delta t \to 0} = P_0^l(t)$$

Therefore, in matrix form, it can be represented as:

$$\begin{bmatrix} P_0'(t) \\ P_1'(t) \end{bmatrix} = [P_0(t) \quad P_1(t)] \cdot \begin{bmatrix} -\lambda & \lambda \\ 0 & 0 \end{bmatrix} \qquad \text{...(3.24)}$$

Assuming that the process starts in state 0 (i.e. at $t = 0$ the system is in UP State) such that $P_o(0) = 1$ and $P_1(0) = 0$; the solution for $P_o(t)$ and $P_1(t)$ may be obtained as below:

$$P_0(t) = e^{-\lambda t} \qquad \text{...(3.25)}$$

$$P_1(t) = 1 - e^{-\lambda t} \qquad \text{...(3.26)}$$

Equations (3.25) and (3.26) are considered as the reliability and unreliability expressions of a single element system (non-repairable). However, it must be noted that such a system or element is characterized by an exponential failure density function.

THE REPAIR/MAINTENANCE CONSIDERATION

The concept of repair/maintenance is associated normally with many equipment/systems. For repair/maintenance money is involved though the amount is a fraction of the money spent during the purchase of the equipment/system. Here, equipment considered is repairable. Markov models have the capability of dealing with repairable equipment/system provided that the failure and repair follow exponential distributions. In case of single repairable system the state diagram is as shown in Fig. 3.6.

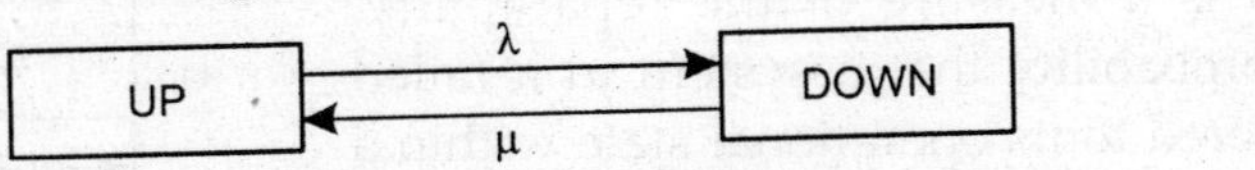

Fig. 3.6: State Space Diagram of a Repairable System

In this case, the differential equation of the system are shown in the matrix form as:

$$\begin{bmatrix} P_0'(t) \\ P_1'(t) \end{bmatrix} = [P_0(t) \quad P_1(t)] \cdot \begin{bmatrix} -\lambda & \lambda \\ \mu & -\mu \end{bmatrix} \qquad \text{...(3.27)}$$

Where μ is the unit repair rate expressed in repairs per year?
Assuming the initial conditions as:

$$P_0(0) = 1.0 \quad \text{and} \quad P_1(o) = 0.0$$

The solution for $P_o(t)$ and $P_1(t)$ are:

$$P_0(t) = \frac{\mu}{\lambda + \mu} + \frac{\lambda e^{-(\lambda+\mu)t}}{\lambda + \mu} \qquad \text{...(3.28)}$$

$$P_1(t) = \frac{\lambda}{\lambda + \mu} + \frac{\lambda e^{-(\lambda+\mu)t}}{\lambda + \mu} \qquad \text{...(3.29)}$$

When t is sufficiently large, the equations (3.28) and (3.29) reduce to those shown below where the probabilities of residing in the UP or DOWN states have been designated respectively as the availability (A) and unavailability (U).

$$A = \frac{\mu}{\lambda + \mu} \qquad \text{...(3.30)}$$

$$U = \frac{\lambda}{\lambda + \mu} \qquad \text{...(3.31)}$$

The limiting values ($t \rightarrow \infty$) are independent of which state the system starts in. The failure and repair process is shown in Fig. 3.7. The value *m* and *r* are often referred to as the mean time to failure (*MTTF*) and mean time to repair (*MTTR*) respectively. The mean cycle time *T* ($T = m + r$) for this process is mean time between failure (*MTBF*). The reciprocal of *T* is designated as the cycle frequency, *f*. For a single repairable component having failure and repair rates (λ and μ) characterized by exponential distributions, result in.

$$MTTF = m = \int_0^\infty R(t)dt = \int_0^\infty e^{-\lambda t} dt = \frac{1}{\lambda} \qquad \text{...(3.32)}$$

$$MTTR = r = \int_0^\infty M(t)dt = \int_0^\infty e^{-\mu t} dt = \frac{1}{\mu} \qquad \text{...(3.33)}$$

In equation (3.33), *M* (*t*) is the maintainability function, which is a measure of the reparability, defined as the probability that a system in a failed state will be restored to its operational state within a specified time period (*t*) when maintenance is initiated subject to stated conditions.

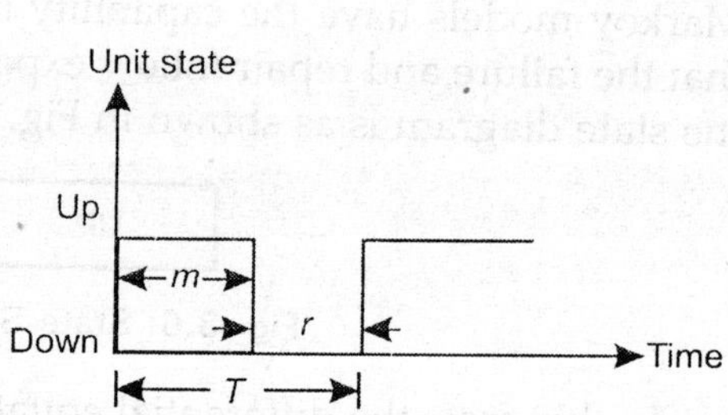

Fig. 3.7: Average Failure and Repair Cycle

Likewise the single element system as discussed above, more complex system structure can also be handled using the Markov model. The complexity of Markov Model depends on the number of system states. In general, a system comprising of *n* distinct elements will have 2^n states, whereas system with n identical elements will have only *n* states. This means that an *n*-element system may at the most need a solution of 2^n first order differential equations. Markov model has the advantage that it can be well extended to system having dependence, repair and standby operation. However, for large number of equations use of computers can be made.

RELIABILITY INDICES

The use of reliability indices correlates the equipment/system performance with respect to time when reliability concept is being used. In order to appreciate the basic reliability indices, the equations (3.30) and (3.31) are to be examined for unit availability and unavailability as follows:

$$A = \frac{\mu}{\lambda + \mu} = \frac{1/r}{\frac{1}{m} + \frac{1}{r}} = \frac{m}{m + r} = \frac{m}{T} = \frac{f}{\lambda} \qquad \text{...(3.34)}$$

$$U = \frac{\lambda}{\lambda+\mu} = \frac{1/m}{\frac{1}{m}+\frac{1}{r}} = \frac{r}{m+r} = \frac{r}{T} = \frac{f}{\mu} \quad ...(3.35)$$

It can be seen from above equations that the frequency of encountering either of the two (UP or DOWN) states in the long run is given by:

$$f = \lambda A = \mu U \quad ...(3.36)$$

In general, the frequency of encountering any given state is given by the product of the probability of being in that state and the rate of departure from it.

In a good repairable system (where $\lambda r << 1$) equations (3.34) and (3.35) can be approximated as:

$$A = \frac{\mu}{\lambda+\mu} = \frac{1/r}{\lambda+1/r} = \frac{1}{\lambda r+1} \approx 1.0 \quad ...(3.37)$$

$$U = \frac{\lambda}{\lambda+\mu} = \frac{\lambda}{\lambda+1/r} = \frac{\lambda r}{\lambda r+1} \approx r \quad ...(3.38)$$

Thus, the three basic reliability indices are λ and r and λr. If λ is the system average outage rate expressed in failure per year and r is the system average outage duration expressed in hours, the product λr is the system average total outage time expressed in hours per year. Interestingly, these indices very well answer the questions, how often and how long, and how many hours per year will the system be interrupted.

System Under Repair Mode

A system consisting of two components in series with outage rates, and repair times r_1 and r_2 respectively has:

System outage rate; $\lambda_s = \lambda_1 + \lambda_2$ ...(3.39)

System average outage duration; $r_s = \dfrac{\lambda_1 r_1 + \lambda_2 r_2}{\lambda_1 + \lambda_2}$ ...(3.40)

System average total outage time; $U_s = \lambda_s r_s$ ...(3.41)

If the two components are connected in parallel then:

System outage rate; $\lambda_p = \lambda_1 \lambda_2 (r_2 + r_2)$ ...(3.42)

System average total outage duration,

$$r_p = \frac{r_1 r_2}{r_1 + r_2} \quad ...(3.43)$$

System average outage time, $U_p = \lambda_p r_p$ ...(3.44)

System Under Maintenance Mode

The component outage due to planned/preventive maintenance is considered as failures. Also, the maintenance outage cannot occur if there is some outage (say breakdown outage) already existing in the system. Assuming the outage rates and outage durations under maintenance of components 1 as λ_1^n, r_1^n and component 2 as λ_2^n, r_2^n the equivalent (combined effect of breakdown repair and preventative/planned maintenance) outage rates and outage durations of component 1 (λ_{e1}, r_{e1}) and component 2 (λ_{e2}, r_{e2}) are:

$$\lambda_{e1} = \lambda_1 + \lambda_1^n \qquad \text{...(3.45)}$$

$$\lambda_{e2} = \lambda_2 + \lambda_2^n \qquad \text{...(3.46)}$$

$$r_{e1} = \frac{\lambda_1 r_1 + \lambda_1^n r_1^n}{\lambda_1 + \lambda_2} \qquad \text{...(3.47)}$$

$$r_{e2} = \frac{\lambda_2 r_2 + \lambda_2^n r_2^n}{\lambda_2 + \lambda_2} \qquad \text{...(3.48)}$$

The outage rate and outage duration of a system having two components with maintenance in series configuration are given by:

$$\lambda_{ms} = \lambda_{e1} + \lambda_{e2} \qquad \text{...(3.49)}$$

$$r_{ms} = \frac{\lambda_{e1} r_{e1} + \lambda_{e2} r_{e2}}{\lambda_{e1} + \lambda_{e2}} \qquad \text{...(3.50)}$$

And for parallel configuration it is expressed as:

$$\lambda_{mp} = \lambda_2 \lambda_{e1} r_{e1} + \lambda_1 \lambda_{e2} r_{e2} \qquad \text{...(3.51)}$$

$$r_{mp} = \lambda_{e1}\lambda_2 \frac{r_{e1}^2 r_2}{\lambda_{mp}(r_{e1} + r_2)} + \lambda_{e2}\lambda_1 \frac{r_{e2}^2 r_1}{\lambda_{mp}(r_{e2} + r_1)} \qquad \text{...(3.52)}$$

A practice that existed over many years to measure the performance of system using the basic reliability indices expressed by equations (3.34) to (3.52) have led to introduce more consistent criterion based on probabilistic considerations for power distribution system reliability evaluation. The new generation of service indices provides a fruitful comparison of the performance statistics collected and to predict the future system performance. The new generation indices used are: System Average Interruption Duration Index (SAIDI), System Average Interruption Frequency Index (SAIFI), Customer Average Interruption Frequency Index (CAIFI), Customer Average Interruption Duration Index (CAIDI) and Average Service Availability Index (ASAI).

The service indices mentioned above can be calculated using the three basic indices described earlier and can be expressed as follows:

$$\text{SAIFI} = \frac{\text{Total No. of customer interruptions}}{\text{Total No. of customers served}} \qquad \text{...(3.53)}$$

$$\text{SAIDI} = \frac{\text{Sum of customer interruption durations}}{\text{Total No. of customers served}} \qquad \text{...(3.54)}$$

$$CAIFI = \frac{\text{Total No. of customers interruptions}}{\text{Total No. of customers interrupted}} \quad ...(3.55)$$

$$CAIDI = \frac{\text{Sum of customer interruption durations}}{\text{Total No. of customers interrupted}} \quad ...(3.56)$$

$$ASAI = \frac{\text{Customer hours of available service}}{\text{Customer hours demanded}} \quad ...(3.57)$$

QUESTIONS

1. With example discuss series reliability model.
2. What is the significance of parallel reliability model in reliability analysis?
3. Discuss the series-parallel reliability model with suitable example.
4. What is the decomposed reliability model? Explain briefly.
5. How redundancy concept can be applied in reliability evaluation? Explain.
6. Discuss different types of redundancy being used in practice.
7. Where Morkov reliability models can be used effectively? Discuss.
8. What is the importance of reliability indices, explain briefly?

4

Failure Statistics for Reliability

INTRODUCTION

The equipment/system, which, are designed and manufactured with lot of precautions and care cannot work continuously for infinite time period. And therefore, failures are bound to occur which will make the equipment completely inoperable or it can work with reduced efficiency. With passage of time elements/components get deteriorated and may fail subsequently if not checked. In reliability analysis, the word failure is used when the equipment/system is not available for use or service. For reliability evaluation, the failure rates of elements/components are essential. Before discussing their details it will be pertinent to know the types of failures, their causes and recording of relevant information. This Chapter will highlight the need of failure data collection, processing and making them suitable for reliability calculations.

FAILURE CLASSIFICATION

Failure classification may be viewed from different aspects according to the effect it will have on the overall performance of the equipment/system. Broadly failures are classified as:

(*i*) System failure, and
(*ii*) Component failure.

In some cases failure of component/element may make the equipment/system completely inoperable and the same cannot be used without repairing the failed component. This is mainly possible in case of high risk equipment/system for example airplane equipment. Even in case of automobiles, the failure of brakes will make it inoperable and therefore, this will fall in the category of system failure. On the other hand failure of a component/element may not make it fully inoperable and the equipment/system can be used with reduced performance. Failure of lights in an automotive vehicle does not make it fully inoperable, the system can be effectively used in daylight. When the elements/components are placed in series, failure of one will make the system completely inoperable. Whereas, when the same are placed in parallel, failure of one element/component may not render it completely inoperable. The system can work but its performance may be reduced. The failure of one of the cylinders in a multi cylinder engine will fall under this category. This type of failure can be classified as component failure.

The engineering classification of failures may have:

(*i*) Intermittent failure, which may result in lack of some function of the component only for a very short period of time, and (*ii*) Permanent failure, where repair/replacement of component will be required to restore the equipment to operational level.

When considering degree of failures, it can be classified as:

(*i*) Complete failure, where equipment/system is inoperative and can not be used further, and

(*ii*) Partial failure, which leads lack of some functions but the equipment/system can be used with care, may be with reduced performance.

Some failures can be sudden and cannot be anticipated in advance, whereas, the gradual failures can be forecast during inspection/testing, which follows the part of the condition monitoring.

Other classification of failure can be:

(*i*) Catastrophic failures, which are both sudden and complete;

(*ii*) Degradation failures, which are both partial and gradual.

CAUSES OF FAILURES

The following are the major causes of failures, which normally occur in engineering equipment/systems:

(*i*) ***Design deficiencies:*** All the performance related parameters couldn't be visualized fully during the design phase. Some of the factors considered during design may not be conducive for the proper functioning of a component and the same may cause failure. These factors can be identified during the performance testing of the equipment/systems. Such factors contribute towards the design deficiencies; however, the same can be incorporated during the redesign phase of the element/component. The various reasons for such deficiencies are the following:

(*a*) It is observed in practice that notch formation in the components/parts create problems due to high concentration of stresses at these points. Such problems are noticed on the machined parts during reduction in their dimensions, the common example being the shaft. Here, the keyways are also subjected to high stress concentration and will be prone to failures.

(*b*) During redesign stage if proper precautions are not taken, due to high stress concentration zones, the components may have premature failures. For example, if the holes drilled in an existing part, may have sharp corners, which will form high concentration regions and will constitute the deficiency in design.

(*c*) During use of equipment, upgrading of parts/components, etc. becomes desirable which may not match with the original design to meet out the new loading conditions of high temperatures, corrosive atmosphere even lack of proper heat treatment. These all again fall in the category of design deficiencies.

(*d*) During the design phase, it is difficult to assess the actual loading conditions or forces to which a part of element will be subjected too, since the same are designed for optimum working conditions. The basic concepts in design include lightweight and minimum cost or sometimes the geometry of the component/element may play an important role. Because of the geometry of the component design deficiencies are very common.

(*ii*) ***Material Selection:*** In actual practice it is generally noticed that maximum failures are due to improper selection of materials, as they do not fulfill the design requirements. The tests, which are carried on the standard material but in actual manufacturing the same may not be used. During design it is believed that the material is homogenous but in real life situation it is not true. Some materials are good either in tension or in compression, but the same may be used where both above conditions do exist. The common example is the beam, which is subjected to tension as well as compression. The imperfection of materials may also cause premature failures because of propagation of cracks in certain areas. Therefore, the selection of proper material for the real working conditions is of paramount importance.

(*iii*) ***Manufacturing Process:*** Dimensional analysis is very important during the design phase to know the stresses to which a particular job will be subjected. In actual manufacturing, it is really difficult to maintain the design dimension tolerances, which again will lead to premature failure conditions. The process of manufacture i.e. machining, welding, grinding, etc. will play important role for the desired performance of the component/element. During manufacture of an item by machining the surface temperatures are likely to vary, which will change the surface characteristics of the material. Some undesired marks, indentation, etc. are the potential source of failure of a component due to formation of cracks during use. Even improper heat treatment may cause surface irregularities which in due course of time may lead to premature failure.

During assembly of parts, due to inaccurate/incomplete specifications may cause early failure of components. Misalignment of the parts resulting in bent shafts, bearings seals, etc. may be one of the primary causes of failure of the assembled parts/components.

The improper service conditions will also affect the rate of failures. Through proper checking, inspection and monitoring the failure rate under such working conditions can be minimized. When newly assembled parts are put in service for the first time, due care should be taken for the unexpected working conditions for example, steep rise in temperature or even slight misalignment. A rapid increase in the speed or sudden rise in pressure may lead to severe damage.

The causes of failures can be many but a few of them are given as under:

(*i*) Wear out failures, where failures are due to normal wear and tear of the parts as expected due to use of the equipment/system.

(*ii*) Misuse failure, these failures can be the cause of over-stressing the component/part beyond their capabilities, and

(*iii*) Inherent weakness failure, these failures are associated during the design and manufacturing of the equipment/system.

In other words the broad classification of the failures can be as follows:

(*a*) ***Service Failure:*** In most of the machinery/system the failures experienced are fracture, excessive deformation and surface failures particularly corrosion. These failures are normally time-dependent and can be checked during inspection and maintenance. Sometimes non-destructive techniques (NDT) may be needed to trace the failures without even dismantling the component.

(*b*) ***Fatigue:*** This type of failure occurs when the loading is cyclic in nature and where cracks initiates and grow, which cause failure of the components/parts. Though the fatigue limits are set in advance but the operating conditions may influence an early fatigue failure. In some Important equipment fatigue meters are attached to monitoring the condition of the system regularly to avoid accidents.

Basically, fatigue starts with the formation of micro-cracks, which may be due to surface roughness or any other reason and are limited to surface grains only. Due to cyclic loading, these micro-cracks propagate and sometimes lead to fatigue failure of the equipment.

The other reason for fatigue can be stress concentration, which may originate from geometrical configurations; however, their values can be determined to avoid failures. Changes in shape may also lead to stress raisers, which may be deliberate due to design requirement or due to manufacturing defects. The more common stress raisers are notches, de-carbonization corrosion, inclusions, internal stress and clamping devices used during manufacture of products.

Surface fatigue can also occur due to heat treatment and rolling loads particularly in ball bearings. Metallurgical changes have limited effect on fatigue life of products. The principle factors are grain size, microstructure and orientation. The vibration in a system may be the prime cause of fatigue failure.

(*c*) ***Excessive Deformation:*** The components/parts are normally subjected to static, dynamic and fluctuating loads. Static loads may be applied gradually or suddenly. Under gradual loads equilibrium of parts is easy to maintain whereas in case of dynamic are fluctuating loading conditions, the components/parts are required to move, which produces inertia forces. If these inertia forces are not properly balanced, will generate severe vibrations and ultimately deformation of the parts/structure.

(*d*) ***Wear:*** Wear is a common phenomena between two surfaces in contact and will cause surface deterioration in the form of micro-cutting, plastic and elastic deformation, surface fatigue, local heating oxidation, etc. With the use of lubricants micro-cracks are excessively pressurized which leads to damage of the surface layers.

The wear in actual form can be classified as under:

(*i*) Abrasive wear, which is caused by ploughing or gauging hard particles against the soft working surfaces. It is quite obvious that the softer material will wear out more than the hard material.

(*ii*) Scuffing can occur when ideal hydrodynamic lubrication cannot be maintained under sliding conditions. Under these conditions, the lubricant layer formed between the matting surfaces break off.

(*iii*) Fatigue wears. This occurs in rolling friction and is caused by fatigue of the surface layers.

(*iv*) Molecular wear occurs when two matting surfaces are subjected to high pressures.

(*v*) Mechanical oxidation wear is caused when the surfaces are exposed to oxidation process in equipment/system.

(*vi*) Cavitations can be caused when surface is acted upon by a fluid flow with local hydraulic impacts.

(*e*) ***Corrosion:*** Corrosion mainly occurs when the equipment/system is subjected to humid conditions and it gradually makes the surface weak and can fail after sometime. The corrosion activity normally depends upon the working environment and can be identified as stress corrosion fatigue or cavitations.

The stress corrosion takes place when stresses are applied on the equipment/system when it is placed in humid atmosphere. This takes place in the form of cracks where material is not even subjected to plastic state. Under fluctuating load conditions where components/parts are placed in corrosive atmosphere, the corrosion fatigue takes place. Whereas the cavitations occurs with the collapse of minute vapour bubbles mainly when the equipment is working in a corrosive environment.

(*f*) ***Blockage:*** When fluid flow takes place in any system, it is likely to get blocked due to formation of sludge's in the fluid. This is mainly found in cooling towers and radiators of internal combustion engines. These blockages create excessive pressure on the surface and cause it to fail.

FAILURE TYPES

Basically failures are defined as the termination of the ability of a component/part to perform its required functions. The failure of component/system can be classified in many ways, which may include the following:

(*i*) Catastrophic failures are ones which immediately stops the working of system/equipment and it can not be used without proper repair/maintenance.

(*ii*) Performance failure. These are related with the performance of the equipment/system. This system may remain operative in the failure of some components/parts but its performance decreases, which is true for the most of equipment used in engineering application.

(*iii*) Deliberate failures are caused either by the neglect of the operating personnel or by his ill intention to make the equipment inoperative for sometime/period. In this case operators make excesses, which are not rational for example application of brakes in an automobile.

Basically, the failures fall under the following categories:

(*i*) Infant or early failures. It can be seen from bath-tub curve, that due to quality of components some equipment, fail during their initial life and such failures number can be high. These failures can also be due to initial turning of the system.

(*ii*) Random failures, which can take place at any time due to unforeseen reasons and it is difficult to predict them. Their causes could be extra stress on the component or the quality of material. However, these failures can be minimized through a proper investigation of load and quality of material in use.

(*iii*) Time-dependent failures where mean time to failures can be predicted since the failures depend on the usage of the equipment/system. Hence, failure distribution can be plotted to know the frequency of failures, which can be used to control the rate of system component failures.

FAILURE INVESTIGATIONS

The technical investigation of failed equipment is sometimes essential to obtain information to know the cause of failure. This information can be utilized only to control further failures but also to bring out some improvement/modifications in the system. However, the correct information can only be had if the investigations are carried out just after the accident/failure and the evidences are not destroyed or distorted. For the same reason investigations are carried out after any major accident to avoid re-occurrence.

FACTORS INFLUENCING FAILURE RATE

The following factors are most important for the failures of equipments/systems and therefore should be given due care in the design and manufacturing phase of the equipment:

(*a*) ***Quality:*** The quality of parts depends on the quality of material used. If the quality of components is higher, will lower the failure rate and will increase the cost of equipment/system. This is more pertinent where reactive materials are used. For example, if acid is to be used, the material should resist its reaction otherwise the component will fail before the scheduled period.

(*b*) ***Temperature:*** The failure rate of electronic components is highly dependent on the temperature of the component. The relationship is expressed as:

$$\lambda = Ke^{(-E/kt)} \qquad \text{...(4.1)}$$

λ = Component failure rate (constant)

T = Component temperature in K

E = Activation energy of failure process (eV)

k = Boltzmann's constant = 863 x 10^{-5} eV/K

K = Constant.

The failure rate depends on activation energy, (i.e. higher the energy, higher is the failure rate). However, the temperature of components/parts also depends upon the surrounding temperature. The electrical insulation properties of materials change at both high and low temperatures. Corrosion also increases with rise in temperature. Even the wear rate of mechanical component also increases with rise in temperature. Viscosity of the fluids also changes with the change in temperature.

(*c*) ***Environment:*** Temperature is one of the examples of environment factors, which affects the failure rate of components. The other environmental factors include, humidity, salt and dust in atmosphere, exposure of frost, nature of process material, vibration, mechanical shock, thermal shock and electromagnetic radiation. Even one environmental factor is enough to cause an early failure but two of these factors may aggravate the failure rate to a greater extent, example is the presence of salt with humidity, which causes corrosion.

The failure rate of instruments in contact with process fluid is significantly higher than those in contact with the background plan environment. Even the cleanliness of the fluid will affect the failure rate.

(*d*) ***Stress:*** In a mechanical system such as beam or strut which are subjected to stresses, produces corresponding strain greater than the elastic limit of the material, the material is likely to undergo plastic deformation and fail. It is, therefore, needed that the strain caused in a material due to application of load should not exceed elastic limit to avoid failure. For this reason only limits are fixed for a load carrying system and factor of safety is used during design stage.

EFFECT OF DESIGN, MANUFACTURING AND ASSEMBLY ON FAILURES

Apart from operational causes for premature failures of component/system, the deficiencies of design, manufacture and assembly can also attribute towards early failures. Faulty design, where proper care has not been exercised for correct loading and other related factors could cause premature failure. However, manufacturing faults causing failures can be minimized through quality control techniques. The analysis of failures would indicate the nature of such failures.

Incorrect assembly of component/system is a critical cause of failure, which can occur during manufacture or even after servicing/repair. Proper care for limits tolerances can reduce the assembly failure.

FAILURE DATA ANALYSIS

For the evaluation of reliability, the information about failures of component/elements is essential, which either can be obtained through experiments or through simulation. Tests are carried out on identical items to study the failure behaviour of the components. Which provide failure data, to give the number of events taking place as a function of time for a particular equipment/system.

Let 'n' be the number of failures that occur in time, $T = t$, then probability of failure in time, $T = t$ is expressed as $F(t)$ and is given by:

$$F(t) = L_t\left(\frac{n}{N}\right) \quad \text{as} \quad n \to \infty \qquad \text{...(4.2)}$$

Where, 'N' is the number of specimens tested. A plot between $F(t)$ and time, T is given below:

Which indicates that $F(T = 0) = 0.$, $F(T = \infty) = 1$. With the assumption, that the N is large. The probability of failure occurring within time $T = t$ is $F(t)$, and the probability of failure occurring within time $T = (t + \Delta t)$ is $F(t + \Delta t)$. Hence, the probability of failure between time, t and $(t + \Delta t)$ is given by:

$$F(t + \Delta t) - F(t) \qquad \text{...(4.2)}$$

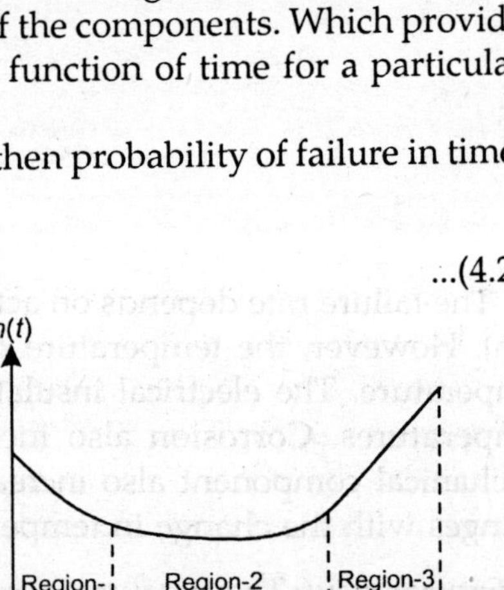

Fig. 4.1: Failure Pattern with Time

The above equation gives probability of failure occurring in Δt t interval at $T = t$. If this equation is divided by Δt, it gives the probability of failure occurring in a unit interval of time at time $T = t$, i.e.

$$\frac{1}{\Delta t}[F(t + \Delta t) - F(t)] \qquad \text{...(4.3)}$$

Where, $\Delta t \to 0$ we get

$$L_t \frac{[\Gamma(l+\Delta l) - \Gamma(l)]}{\Delta t} = \frac{uF(l)}{dt} \qquad ...(4.4)$$

This equation represents the failure density function $f(t)$ and is the slope of $F(t)$ curve at function of time:

$$f(t) = \frac{dF(t)}{dt} \qquad ...(4.5)$$

FAILURE RATE AND ITS RELATION TO RELIABILITY

The failure rate is defined as the probability that a failure per unit time occurs in the interval (t_1, t_2) given that a failure has not occurred prior to the beginning of the interval t_1, the interval failure rate (FR) is expressed as:

$$FR(t_1, t_2) = \left(\frac{1}{t_2 - t_1}\right) \frac{R(t_1) - R(t_2)}{R(t)} \qquad ...(4.6)$$

Where, $R(t)$ is the reliability at a given time period. When the time interval $(t_2 - t_1)$ approaches zero the failure rate is called as the instantaneous failure rate. It is also termed as hazard rate or hazard function and can be expressed as:

$$h(t) = \frac{1}{R(t)} \frac{dR(t)}{dt}$$

Which be expressed as:

$$h(t) = \frac{f(t)}{R(t)}$$

Where $f(t)$ is the probability density function.

The difference between $h(t)$ and $f(t)$ is that $h(t)$ is the rate of change of conditional probability of failure given survival time 't'. Whereas, $f(t)$ is rate of change of unconditional probability of failure, the hazard rate indicates the change in the failure rate over the lifetime of the component/ element. If $h(t)$ is increasing in $t > 0$, then $f(t)$ is said to be increasing failure rate and vice-versa.

HAZARD RATE

The typical graph showing hazard rate is called as bathtub curve. This curve shows three distinct regions being 1, 2 and 3. The region 1 shows pre-mature failures due to material or manufacturing defects, which can be minimized by better quality control. The second region is termed as random failure with horizontal curve indicating that the failures are constant in nature and may be due to fluctuating stresses under operating conditions. For these failures due care is taken in design phase. The third region is termed as wear out failure zone where failures are due to fatigue deterioration and stresses, etc. and are caused due to prolonged use of the equipment/system. The failure rate under this region increases rapidly. The figure given below is for electrical/electronics components which relatively have longer useful life period. However for the mechanical components the curve is slightly different and useful life period is short.

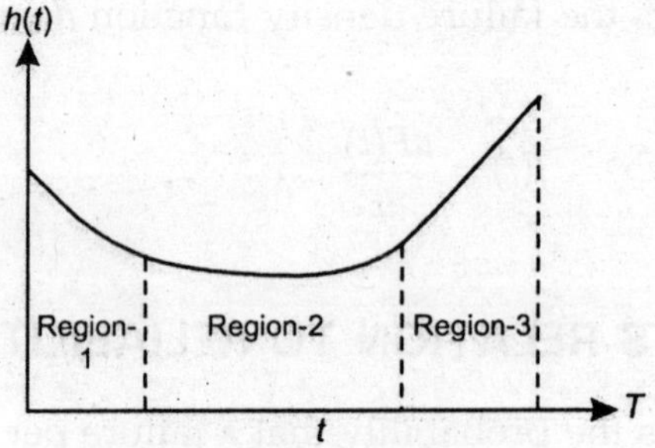

Fig. 4.2: Hazard Rate

HUMAN FACTORS IN EQUIPMENT FAILURES

Mechanical failures can be known in advance with its reasons. However, the human factor is highly important factor in accidents and is well known to the organizations involved in insurance activities for example automobile industries. Studies carried out in this area have highlighted the human factors needing attention during the accident. The attention should be given to the effects of 'boredom' group interaction, lack of instructions, incorrect motivation, ill defined area of responsibility, poor control, poor communications and many similar causes of irrational behaviour.

Fault Detection Sensors

To identify the faults during scheduled maintenance or check-up to examine physical or chemical change, various sensing devices are being used and include X-ray fluorescence, corrosion meters, load cells, transducers and strain gauges etc.

QUESTIONS

1. How failures are classified? Explain it briefly.
2. What are the main reasons for failure of system? Discuss.
3. Explain, how failures can be investigated in components/systems.
4. Highlight the influencing factors associated with failures.
5. With the help of an example explain data failure analysis.
6. What is the significance of hazard rate in mechanical systems?
7. Write the importance of human factors in equipment failures.

5

Design for Reliability

INTRODUCTION

In earlier Chapters, it is indicated that design of product/system plays an important role to improve the reliability. Here, the objective is to design an equipment/system, which meets the target failure rate under actual working conditions. It is seen that the failure rate will depend on temperature, environmental conditions and stresses to which the said equipment is subjected to. Through proper analysis lower and upper specifications limits for failure rates can be identified and if equipment/system failure rates are within the limits as specified, it can be accepted otherwise it will be rejected.

The main requirement of design for reliability is that independent checks are carried out to ensure that the 'paper' design meets the specification. A team of experts along with design group can ensure that the affects of all likely variations in environmental conditions and component failure rates have been considered. This process concludes with the design review board, which makes the final decision to approve the design.

The project team after approval can make proto-type design. Here number of proto-types are made and tested under the range of specified environmental conditions to ensure that the above specification is satisfied then only the final product will be accepted.

DESIGN PRINCIPLES FOR RELIABILITY

The basic objective of design is that the equipment/system meets the target failure rate (or expected reliability) under the given conditions. Normally failure rates are considered constant with time to make the reliability model simple. However, the following principles of design for reliability be considered:

(*i*) Selection of components/parts
(*ii*) Functional rating of all parts
(*iii*) Environmental conditions with their impact on equipment/product
(*iv*) Complexity of equipment/system
(*v*) Redundancy/diversity
(*vi*) Reliability estimation

The product therefore should be designed so that the failure rate is minimum i.e. the failure free period is maximum. The techniques of Failure Tree Analysis (FTA) and Failure Mode and Effect Analysis (FMEA) can be used. The same have been discussed in the later part of the chapter.

The Design Process

The ideal process of design starts from a problem to solution with the following stages:

(*i*) Requirement—Problem definition with details for which solution is desired.
(*ii*) Specifications—List down the additional constraints which influence the performance of the product/system to obtain solution.
(*iii*) Basic design—It is the best overall solution possible under the specified constraints.
(*iv*) Detail design—It is the refinement of the solution with respect to the points required for the manufacture of equipment/system.
(*v*) Development—Improvement of the obtained design which is now possible through computer aided design (CAD).

To eliminate the design, continuous performance characteristics (x) will depend on both design parameters $\{p\}$ and random variables $\{r\}$ are to be examined.

$$\{x\} = f[\{p\}, \{r\}]$$

A model for robust design can be developed based on the above parameters using computer simulation. Designer must specify the random variable input vector {r}, its significant effect $\{x\}$ and the likely range for each variable. The next stage is a parameter design experiments, which is defined by two matrices, the design parameter matrix and the random variable matrix.

Optimal design for the user should maximize product popularity and the number of units sold. It should also minimize the user's loss. Attention should be given to the design process, to manufacture of the product besides the product design. Robust methods can also be used in process design as well.

Deficiencies in Design

When due attention is not focused to the important factors responsible for proper functioning of the parts, failures result. While cases of such neglect may be examined carefully, will indicate design deficiencies with the following reasons.

In mechanical components the presence of notches causes point of high stress. The change of cross-section in shafts also has high stress concentration, which can be part of major deficiencies in design and must therefore be avoided.

DESIGN TEST

A number of items of a given product should be placed on test, under a given set of temperature, stress and environmental conditions, for a total observation time T and total number of failures N_F during time T are recorded. The observed failure rate ($\bar{\lambda}$) is defined as the number of failures divided by total up time i.e.

$$\bar{\lambda} = \frac{N_F}{\text{total up time}}$$

The test should be repeated at different sets of relevant temperature, stress and environmental conditions at the set of observed failure rates ($\bar{\lambda}$) and compared with the specification range i.e., lower specification limit and upper specification limit.

Once it has been proved that the prototype design meets the specifications, the next stage is usually to build product models. Production models can also be tested over the full range of environmental conditions and field limits, in the normal working environment. The above tests carried out should provide the following information on:

(*a*) The basic individual causes of failure, and the component or the conditions responsible for the same.

(*b*) Number of failures due to each basic cause as a percentage of the total number of overall failures.

Using these data plot can be drawn showing the number of failures with the causes of failure. This information will indicate the basic causes of failure for remedial action. Here emphasis is to be given to find out the actual cause of the failures i.e. design or material or process etc. which ultimately will make the selection of the above parameters.

Test for Design Evaluation

The following tests can be conducted to evaluate the design:

1. Performance—Check that the product meets the performance specifications as desired.
2. Environmental condition—For the same product the working environment may be different, so examine that the product meets all above conditions.
3. Stress level—Where the products are subjected to loading (mechanical) it must be examined critically that it can withstand all type of specified loading.
4. Reliability—Check that the product meets all reliability requirements as stated.
5. Maintainability—Examine that the time required to repair/maintain a product is minimum and it is easy to carry out work.
6. Life—Through tests examine the working life of the product without under failures.
7. Trial-run—During the test run check that the manufacturing used are capable of meeting the design requirements effectively, also check the reliability.

GENERAL PRINCIPLES OF DESIGN FOR RELIABILITY

The following general principles of design should be observed:

(*a*) *Component/element selection:* The components to be used should be well within the failure data limit. The technologies, which are well established, should be used particularly in the area of electrical/electronics visualizing the reliability factor of one over the other.

(*b*) *Factor of safety:* Which is an important parameter for the design of components/equipment where loads/stresses are of varying nature. A system is likely to have premature failures if subjected to over loading. For mechanical element/component a safety margin of better than 5.0 should be used, and in case an electronic circuit the voltage stress rate should be kept well below 0.7 to minimize their failures.

(*c*) *Environment:* In earlier sections, it is pointed out that the failure rates are critically dependent on the environment. Therefore, environmental factor must be considered and elements/components with a high quality level, which are capable of withstanding the environmental conditions only, should be used. Such components will have low failure rate and will compensate for high capital cost involved in their production, if any.

(*d*) *Complexity of System:* It is seen that in case of series reliability model, the failure rate is added depending upon the number of components used. Thus, the number of components/

elements in the system be as minimum as possible to perform its functions. In electronic systems reliability can be improved by using integrated circuits, which replace many hundreds or thousands of basic devices. The failure rate of integrated circuit is generally less than the sum of the failure rate of the devices it replaces.

(*e*) *Redundancy:* It is noticed that using parallel system, increases the reliability of the overall system. This should only be considered where certain components/elements have high failure rates since it adds to cost.

(*f*) *Diversity:* In a system where failure of one component causes failures of other components since all share common power supply. Failure of power supply thus causes, all of the circuits to fail. If the probability of common mode failure limits the reliability of the overall system, the use of equipment diversity should be considered. Here, a given function is carried out by two systems in parallel but each system is made up of different elements with different operating principles. The example is a temperature measurement system using electronic and pneumatic systems, which may have different failure pattern.

(*g*) *Reliability Calculation:* From failure rates of the components/elements the overall system reliability can be calculated depending upon their configuration i.e. series, parallel and series-parallel as the case may be. The calculated reliability/failure rate for the overall product should then be compared with the target value, if this value is not found within the specified limits, the design should be adjusted until the target figure is reached.

FAILURE ANALYSIS FOR DESIGN

In many cases failures of components occur only in one way however, for some components/ elements it can occur in two/many ways. For example, an electronic component such as a transistor or diode may fail, open circuit (maximum resistance) or short circuit (maximum resistance) i.e. has at least two failure modes. In more complex circuits, the open circuit failure of a given component will cause a certain element failure mode and short circuit failure of that component will cause a different failure mode.

Let us assume that open circuit failure of a given component causes high output and short circuit failure causes the element output to fall to a low value. We further assume that above system controls the fluid flow in a pipe. If an element fails to high value then the controller closes the valve to reduce the flow and if the element fails to a low value then the controller closes the valve to reduce the flow. Two failure modes therefore, result in two different control loop failure modes. Depending upon the circumstances, failure mode could be fail-safe failure, where the plant is moved to a safe condition and one mode will correspond to a fail danger failure where plant is moved to dangerous conditions. It will depend upon the type of fluid (colder/hot).

The system/equipment should be designed to take care so that the fail danger probability is minimized. Three main techniques available to the designer are (*i*) fault tree analysis, and (*ii*) failure mode and effect analysis, and (*iii*) Event Tree Analysis. Which are discussed briefly. These must be used to improve the equipment/system design to minimize the failure rates.

(*a*) ***Fault Tree Analysis (FTA):*** It is a systematic way of identifying faults whatever possible to lead a system/equipment fail-danger system. In this case, a logic diagram is constructed

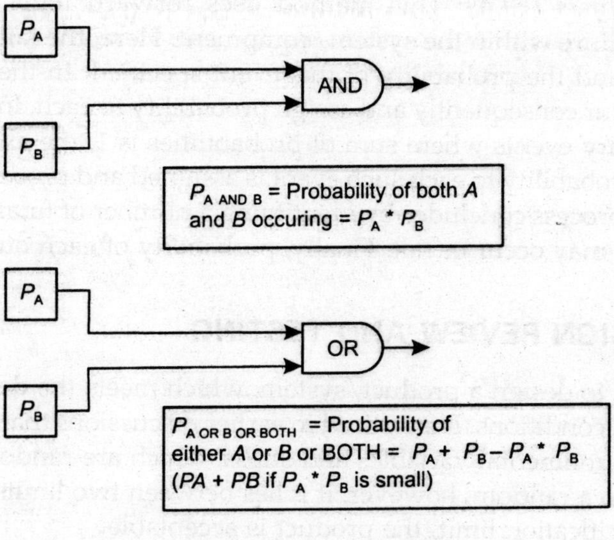

Fig.5.1: Fault Tree Analysis

showing all possible combinations of faults and conditions that could stop the event. This diagram is prepared within number of AND and OR gates. The figure shows the methods of combining probabilities.

In a simple example of steam power plant, if the flow rate of steam to the boiler goes high, may cause failure of vessel. Here, the temperature control loop consists of a platinum resistance thermometer, a transmitter, a controller, a current to pneumatic converter and a control valve.

The control valve is shut at a minimum pressure and fully opened at maximum pressure. The plant is safe guarded by a pressure trip system consisting of a pressure switch and a three-way solenoid valve located in the airline between converter and the control valve.

Under normal conditions, the pressure switch is closed, the solenoid valve is energized, the vent port shut and the air supply to the control valve is normal. If the column pressure exceeds the trip setting, the pressure switch opens to de-energize the solenoid and the vent port opens, this makes the control valve to stop the supply, thus no steam is supplied and makes the plant safe. The trip system, therefore, increases the probability that any failure will be a fail-safe failure. Still opening of control valve may cause high flow rate and the probability of failure remains, which can be calculated by identifying all possible cause of event, with the help of fault tree diagram to get overall fail danger probability.

*(b) **Failure Mode and Effect Analysis (FMEA):*** In case of large number of components, FTA may not be practicable and failure mode and effect analysis may be more useful. Here, all significant component failure modes are listed and their effect on element i.e. whether fail safe or fail danger, evaluated and a failure rate assigned to each failure mode. This gives the sum of the component fail danger/fail-safe rates, which, can be used to calculate the two probabilities. Having known the failure rates component/system can be designed to meet the above requirements.

(c) ***Event Tree Analysis (ETA):*** This method uses forward logic to define all possible consequences of a failure within the system/equipment. Here, the failure is first identified as the initiating event and the probability of this event specified. In the next step, identify the event, which will occur consequently and assign probability to each. In many cases, there may be two complementary events where sum of probabilities is 1, the example is the alarm that works or does not. Probability to each such event is assigned and process is continued until the line is built up. The process concludes by identifying a number of final outcomes, for example whether an accident may occur or not. Finally, probability of each outcome can be found.

RELIABILITY DESIGN REVIEW AND TESTING

The objective here is to design a product/system, which meets the desired failure rate under given environmental conditions. It is noticed in earlier discussions that failure mainly depends on temperature, environmental variables and stress, which are random in nature. Therefore, the failure rate is also a random, however, if it lies between two limits i.e. lower specification limit and upper specification limit, the product is acceptable.

The main requirement of the reliability design review is that there are independent checks to ensure that the 'paper' design meets the above specification. This can be done properly when design team works with the review team. The final decision of the review team decides the acceptance/rejection of the product.

In the next step, paper design is converted into a prototype. Here, number of prototypes are made and tested under varying environmental conditions to satisfy the design specifications. A statistical analysis can be carried out to ascertain the final acceptability of the product.

A designed product/system will be subjected to two conditions i.e. achieved performance and required performance. Their correlation will indicate the reliability level. If the achieved performance matches the required performance, the reliability is one and the product is most reliable. The variability in performance indicates the un-reliability. Therefore, reliability is always function of required performance and achieved performance.

$$R = f(Q, H)$$

Where, Q = complete required performance function

H = complete achieved performance functions.

The information on the above can also help the designer to improve the reliability of the product. Because it is really difficult to achieve the required performance from any equipment/ system.

Role of Computer Aided Design (CAD)

The concept of Computer Aided Design (CAD) and Computer Aided Manufacturing (CAM) have brought revolutionary changes in design. With the conventional design procedure it used to take years to develop, design and produce a product. Now with the use of CAD, it is possible to design a product in a shortest possible time with many alternative solutions. All the design parameters can be varied and their influence on the product can be studied without actually manufacturing it. The techniques like Finite Element Analysis (FEM) can be applied to take care of different type of loading to determine the failure mode and to opt for alternatives. Computer aided manufacturing (CAM) is also helping the design engineer to incorporate desired changes in the product to make it more reliable. Further, two-dimensional and three-dimensional modeling helps to visualize the product on the screen to judge its appearance.

QUESTIONS

1. Explain, what design tests can be carried out to predict failures.
2. Enlist the general principles of design for reliability.
3. Discuss how failure analysis helps during design phase.
4. Explain fault tree analysis with suitable example.
5. What is failure mode and effect analysis? Explain it.
6. How event tree analysis helps during design phase?

6

Quality and Reliability

INTRODUCTION

The quality of a product/system can be defined as its ability to ensure complete customer satisfaction in totality, which may include all aspects such as comfort, appearance etc. For example, in case of automobiles such parameters are given due weight-age. However, the quality of engineering products can be measured in terms of its characteristics that contribute overall performance, which satisfies a customer/user's requirements. These are termed as performance characteristics, which are of three types i.e. continuous, discrete and binary.

Continuous characteristics can take any value within a given range, for example, fuel consumption, which can take any value between 10 and 60 km per litre. The discrete characteristics are related with appearance, comfort, etc. and cannot be quantified which are subjective in nature. On the other hand, binary characteristics yield result in terms of two conditions. The said objective is either acceptable or not i.e. in terms of 0 and 1. In general the products can be specified using continuous performance characteristics. Their values can be established by independent measurement of performance identifying characteristics.

All engineering products are subjected to random effects in the components, which make up the product. The physical dimensions of any two products cannot be exactly the same. However, environment plays an important role. These random effects can be regarded as random inputs to the product, which result in corresponding random variations in the output, of continuous performance characteristics of the product. This random variable may be the source of dissatisfaction to the customer. For the very reason the specification limits are specified for almost all products.

QUALITY CONTROL OF PRODUCTS

Visualizing the random variations in product characteristics, it is required to adopt the quality control of the product. Some statistical methods are employed using upper and lower specification limits to monitor the quality. Here concept of mean and standard deviation is used to either accept or reject a lot, which is being tested. The whole exercise is termed as statistical quality control (SQC). There are two main sources, which are responsible for any variations in product characteristics, which include the following.

(*a*) The variability brought-in the components due to design defects and due to poor quality of the materials.

(*b*) The variability of the production process may also affect the quality of the products i.e. use of different machine for the same job or use of different processes for the same product.

SQC techniques can be used to identify and limit both above sources of variability. When this technique is applied to any process, it is termed as statistical process control (SPC). Here, line measurements of product performance characteristics are made and compared with the specification limits. This will not only indicate percentage of error, off-specification of product but should also identify the sources of variability in the process, which can be monitored and minimized after being located.

PRINCIPLES OF QUALITY

Quality is associated with the entire product, which are being manufactured by any process. The cost of the product is directly proportional to its quality. For ultimate satisfaction of the users/customers, the concept of quality has to start from the inception of the product. If the raw material used does not posses good quality, the final product cannot meet the required quality standards. Even if the quality of the raw material is satisfactory, the process of manufacture will also contribute for the products quality. To check the quality of the products standards are set by the government and the products, which pass through the tests conducted by such organization, acquire the standard make, which ultimately helps in selling the product. For this purpose Government of India, has established the Indian standard. Therefore a particular product dominates the market due to its quality only.

The quality of some products cannot be measured in terms of comfort it provides to the users. For example, the quality of a room in a hotel can be judged by the facilities and the comfort derived by the user. For the engineering product, the quality can be quantified i.e. it can be measured. The quality of an automobile can be expressed in terms of performance characteristics i.e. mileage it gives besides the comfort it provides to the users.

STATISTICAL QUALITY CONTROL

For measurement of quality, predefined standards are set with upper and lower limits. If a product's performance falls within the specified limits, it is accepted, otherwise, rejected. Depending upon the products, inspection has to be carried out either on each item or on a sample drawn from the lot. When sample inspection is carried out, the lot is either accepted or rejected depending upon it's meeting out the quality specification within the said standard limits. This system is normally followed for mass production since inspection is not possible on each item/product. Whereas, in case small number of products, the inspection has to be carried on all items individually, for its acceptance or rejection.

For the purpose of continuous performance evaluation of different parameters of a product/equipment, is defined, with target values. However, all engineering products are subject to random effects in the components, (which make up the product), in the manufacturing process and in the environment where a particular equipment/product is to be used. The variations take place around the target values with upper and lower limits, which are called as specification limits.

To know how the actual random variations in a given performance characteristic say x, for a given product, compare with the target value x_T, within the given upper and lower specification limits. A statistical analysis for these variations is required and involves calculation of mean standard deviation and probability density function of the variations.

To carry out the exercise assume that N samples have been taken from the total population of values of x. These are given as x_i , where $i = 1, 2 \ldots. N$

The mean and standard deviation can be expressed as given below:

$$\bar{x} = \frac{1}{N} \sum_{i=1}^{i=N} x_i \quad \text{...(6.1)}$$

$$S = \sqrt{\frac{1}{N} \sum_{i=1}^{i=N} (x_i - \bar{x})} \quad \text{...(6.2)}$$

The values of mean and standard deviation specify the center and the spread in the sample values but do not give information about the shape of the variation, which may have any profile. However, this information can be obtained by probability and probability density functions. The probability can be expressed as the ratio of number of occurrence of the event to the total number of trials, when these are made independently. Whereas probability density function can be expressed as:

$$p(x) = \lim_{\delta x \to 0} p_j$$

Where, J specifies, the section for $j = 1, 2 \ldots. m$ and δx is the width of each section.

After comparing the statistical distribution of each characteristic with given specification limits, it is to be decided to accept or reject the product.

THE CONCEPT OF TOTAL QUALITY

The failure of a product/system either causes direct financial loss or inconvenience to the customers/users. Examples could be late running of public transport system or power failure in a market complex. In engineering context, a defective machine selling may lead to defective components and rework of the same with additional costs and time delay. Since parts may get damaged during transportation due to incorrect packaging.

Under all circumstances poor quality may be defined as a failure of product or service to perform intended operation successfully, which ultimately does not satisfy the customers/ users.

Taguchi suggested that the deviation of a product from an optimum performance might be described by a "quality loss function" which can be selected with monetary value. The objective of an organization should be to minimize the quality loss function of its product/ service. Under total quality function all parts of the organization are responsible for its success/ failure. The policy of the organization should be of continuous improvement of its products/ services.

The key principles of total quality are following:

- Adopt policy of continuous improvement of all areas, including training in the newer technical fields.
- Reduce number of suppliers, and try to involve them in continuous improvement also.
- Provide on-line techniques to identify problems and their solutions using statistical methods.

- Make use of multi-disciplinary teams, with open and congenial environment for effective participation of all working groups.

The following quality engineering techniques can be applied to improve the quality of product/service:

(*i*) Quality Function Deployment (QFD) is a matrix technique for identifying product/process design requirements and efforts required in meeting these requirements.

(*ii*) Design for Manufacture and Assembly (DFM/DFA). This describes methods that assist in designing a product for ease of manufacture and assembly.

(*iii*) Failure Mode and Effect Analysis (FMEA) is a systematic approach for identifying possible modes of failures of a product or process and their consequences on such failures.

(*iv*) Taguchi Methods is a statistical method to evaluate quality loss function in order to minimize variation in design performance.

(*v*) *Poke Yoke:* It is the term used to describe proof devices within a process to prevent defective products being produced. For example, the shape of a part is such that it can be fitted in only way to avoid confusion to the technician.

(*vi*) Process capability studies involve about correct information in connection with required tolerances for a particular process.

(*vii*) *Simulation:* This can be used where past data/information on the subject matter is not avoidable.

(*viii*) *Validation and testing:* This involves a number of tests methods for verifying that a product design performs as expected.

(*ix*) Besides above all, problems solving techniques can be used for identifying of a problem and its solution.

STATISTICAL PROCESS CONTROL

The quality of a product/system can be measured using its performance characteristics. If these characteristics are variable in nature can be represented by statistical distributions. The quality of the product can be checked against the laid down target values within pre-specified upper/lower specification limits for each parameter. When such parameters have process involvement, the quality control is referred to as 'statistical process control'. Here, on-line inspections/tests are carried out and compared with standard specifications within the limits. This can identify sources of variability in the process, which can be corrected and modified, if required.

It is essential to know that a particular process is capable of producing a product within the specified limits. Here, capability indices, which are decided in advance, can be compared with the standard deviation of product characteristics, caused by process variability.

QUALITY CONTROL CHARTS

For mass production of products, it is difficult to measure the performance of each component individually, since it involves cost and time. In such cases, samples can be drawn for inspection/testing and their probability density function values are calculated to compare its distribution with the mean values with upper and lower limits. Charts can be prepared for the samples drawn from the populations. These charts can help the users to accept or reject subgroups of drawn samples. The mean and standard deviations are calculated for each sub-group before making a comparison. This helps in identifying a defective sub-group only, a good amount of time is saved in carrying out inspection/testing of entire population.

THE ACHIEVEMENT OF QUALITY

As indicated in earlier discussions, quality of a product/system starts from its raw material to the finished product. To achieve high quality of any product it is essential that at a first stage, the raw material must be of desired quality. Defects do occur during manufacture of product/system, it is therefore, necessary to check the process through which a product has to undergo. Inspection/testing also play important role in manufacturing of good quality products. So, due care should be taken during such activities. Proper handling of product is also essential to maintain its quality at the user's end. Sometimes it is observed that many products/items get damaged during their transaction. Even during installation of equipment, care has to be exercised to maintain its quality. Products/systems can only retain their quality with the proper support of all individuals associated with it. At this point of time it should be noticed that the quality of the product in use is also important which, can be maintained by it's proper use.

QUALITY MANAGEMENT

For achieving the desired quality of a product an effective quality management is essential, which can be better managed by a quality manager. The responsibilities of a good quality manager include the following:

(*i*) Setting quality objectives i.e. what level of quality is required for the product since it involves cost.

(*ii*) The formulation of quality system to achieve the objectives set by the organizations.

(*iii*) Implementation of the quality system for effective results, which needs cooperation of all associated persons.

(*iv*) Preparation of quality manuals for giving the details of quality system to be practiced in the organisation.

(*v*) Training and education of the personnel who are involved in the design and production systems.

(*vi*) Information recording should be done properly to know the deviations from the set norms.

(*vii*) Proper testing/inspection of the product/equipment with the help of authentic testing devices.

(*viii*) To take corrective action as and when required, to achieve desired quality.

(*ix*) Quality system audit is also essential to know its implementation at all stages of production.

(*x*) Financial benefits associated with quality system, must be calculated to justify its application in any organisation.

IMPACT OF QUALITY ON RELIABILITY

It is observed from failure analysis of equipment that good quality products/components have less number of failures as compared to poor quality parts. If proper attention is paid to the quality of products their failures also will be minimized. The quality not only improves reliability of equipment/system but also aids to its value, which is very vital in today's technologically developed nations. While evaluating the overall reliability of equipment, the reliability of each part/component is taken into account. Therefore, the impact of quality on reliability is very vital and significant. The quality products also draw attention of the users/customers, as the maintenance problems associated with such equipment/services are also minimised.

QUALITY DESIGN REVIEW

Independent checks must be carried out to ensure that the design meets the quality specifications as defined by the parameter design variables. The review team must closely work with design team for suggestions for improvement at appropriate stages in design. All design changes resulting from this review process should be controlled and documented. Again there should be independent checks that any redesign also meets the quality specifications.

In the next step the paper design should be changed to prototypes. Here a number of prototypes, each with different design parameter are built, each one is tested over a specified set of values of random input variables. Here again there should be independent checks to ensure that the chosen prototype design meets the quality specifications.

The random variables for the process will be made up of the following:

(*i*) Random variables in input components and material. The design parameter variables for product can be used to identify these random variations.

(*ii*) The random variations in the process itself.

QUALITY IN MANUFACTURE

The quality of the product mainly depends on the input variable such as material of the product and its manufacture process. To control the above two variables the concept of quality control must be applied to maintain the specification limits. The test procedures to be used should be agreed between component/material suppliers and the customers. Sampling techniques are essential if there are large number of components/parts, again this must be agreed by supplier and customer.

All manufacturing process are subject to random variations because of human factors. To eliminate the human factor the use of automatic/computer controlled machines/equipment can be used. Sometimes, automatic processes also show variability mainly due to temperature variations.

QUALITY TESTS

The quality tests are required to show that the finished product (*i*) meets the specification with desired performance and (*ii*) also meets specification for continuous performance characteristics.

QUESTIONS

1. Discuss the role of quality associated with reliability.
2. How quality control of products can be done, discuss briefly?
3. Enlist the basic principles of quality in reference with reliability.
4. What is statistical quality control? Explain it.
5. Discuss how the concept of total quality can be applied to engineering products.
6. Explain how quality can be achieved in practice.
7. Discuss the need of quality management.

7

Reliability and Simulation

INTRODUCTION

The concept of reliability and its evaluation have been discussed in detail in the previous Chapters. In this chapter presentation is based on that, how simulation can be used in reliability assessment. The basic reasons for this are twofold. Firstly, it is difficult to procure failure statistics from the field with required level of confidence, where equipment/machines are deployed. Secondly, authenticity and accuracy of the information collected, is questionable, which may not be correct in all respect. Therefore, simulation studies are necessary to examine the feasibility of various reliability models used in practice for analyzing the system's performance. Also, the effects of the changing parameters in a model upon the system behaviour could be analyzed using simulated responses of the pertinent factors guiding the overall performance of the equipment/system under consideration. Through simulation, an investigation can be made to establish suitable performance indices to represent the functional behaviour of the system under all operating conditions.

Simulation is a quantitative technique, which describes a process by developing a model of that process and then conducting a series of organized experiments to predict the behaviour of the process over a given time period. Observing the experiments is very much like observing the process in operation. To find how the real process would react to certain changes, we can produce these changes in our model and simulate the reactions of real process which are to take place in actual working environment.

For the solution of the real life problems, scale model can be build and its behaviour can be observed under simulated conditions, which must be very close to the real life situation. In simulation, we build a mathematical model, which cannot be solved, and run on sample data to simulate the behaviour of the system. Thus, simulation involves performing experiments on the model of the system. The conclusions drawn can be useful to take precautionary measures necessary for minimization of downtime of the equipment/systems. For the purpose, computer program have to be developed with some special capabilities and flexibilities for allowing the detailed study of equipment failure/repair statistics for predicting MTTF and MTTR, to make reliability calculations.

Simulation is an appropriate substitute for mathematical evaluation of a model in many situations. Although it involves assumption, but they are manageable. The use of simulation enables us to provide in-sight into certain management problems where mathematical evaluation of a model is not possible.

RELIABILITY BASED PROBLEMS

Following reasons can be given, why simulation can be used to solve the maintenance related problems:

- Simulation may be the only method available because it is difficult to observe the actual working environment of the equipment.
- Practical difficulty in developing mathematical models of complex situations arising during the study.
- Actual observation of a system is difficult and time consuming, since the information is required over a long period of time to make accurate policy decisions.
- Failure predictions are random in nature and cannot wait for the failure to occur in real life, for collecting information. When the equipment is being used for the first time the information required of the past is practically impossible. Use of simulation in place of other techniques and models, involves a trade off because of the following limitations and demerits.
    - Simulation is not precise. It is not optimization and does not yield an answer but merely provides a set of system's responses to different operating conditions. In many cases, this lack of precision is difficult to measure.
    - A good simulation model may be expensive. Often it takes a very long time to develop usable corporate planning model.
    - All situations cannot be evaluated using simulation, only situations involving uncertainty are candidates, and without a random component, all simulated experiments would produce the same answer.
    - Simulation generates a way of evaluating solutions but does not generate solution by themselves. Users must generate the solutions, which they want to test.

SIMULATION PROCEDURE

All effective simulations require a great deal of planning and organization. Simulation, although vary from situation to situation but in general, the following steps may be followed which are common in nature:

- Definition of the problem with clear understanding to formulate the model.
- Formulation of the model based on the real life situation considering all possible conditions.
- Testing of the model, its comparison with the actual problem.
- Collection of relevant information to test the model.
- Run the simulation, wait till it stabilizes, so that the result obtained are realistic in nature.
- Analyze the results of the simulation and if desired, change the solution you are evaluating.
- Validate the simulation, this involves increasing the changes of the inferences you may draw about the real situation from running the simulation to become valid.

DATA PROCUREMENT

It is seen in practice that the users of the equipment/system, do not make sufficient quantity of information on failure prediction, etc. available to maintenance department to carry out investigations for the following reasons:

- Information received from user agency may not be adequate to conclude failure pattern.
- Data have not been collected from actual system but from similar type of equipment, which may not yield the correct results.
- The operating conditions must have not matched with the actual working conditions.
- Environmental factors have not been given due recognition, since their effect on the system is vital.

The following factors must be given due consideration while using the field information:

- Similar or identical components are used under different working conditions and under different maintenance policy.
- The failure pattern of parts/components may have different effect upon the performance of equipment/systems.
- The design, manufacturing/fabrication of different parts may vary, therefore, may affect their performance.
- The population of the sample size may be small for reliability evaluation and the results obtained may not match the actual working conditions.
- The failure pattern of mechanical components may be different from place to place and from one working condition to another.

DESIGN OF SIMULATION MODELS

The results of simulation model basically depend upon its model design, if the simulation modeler overlooks the requisites of simulation that have previously been discussed, the study may fall far short of meeting its objectives or may indicate false achievements when there is none. Likewise, if the model fails to take into consideration the important aspects of the real world system, then the model cannot be validated and system behaviour based on the model, may be erroneous. During design stage, two aspects can be considered firstly, simulation technique and secondly practical restrictions.

The design related problems could be minimized with the use of available simulation languages. The parameters required during the design stage include, time control, random number generation, stochastic variate generation, variance reduction techniques and simulation languages.

The time control is divided in two groups i.e. uniform time flow and variable time flow. Under uniform time flow, model is advanced and processed at a fixed time interval whereas, in the case of variable time flow, the model moves according to the movement of events which is preferred when the process occur at irregular time interval. The fixed time mode is useful when occurrence of events is regular in nature.

Simulation is useful tool because of its ability to handle complex systems that require modeling of interacting stochastic variates. These are the means for modeling empirical or theoretical distributions of real world parameters. Stochastic treatment of relevant events greatly increases the realism and applicability of the simulation model. These stochastic variates are generated by manipulation of pseudo random numbers calculated by the computer. Variance reduction methods when used in the model can reduce the time and cost of simulation by minimizing the number of interactions to bring the model to study state.

Some of the general-purpose simulation languages are SIMSCRIPT 2.5, GASP II/IV DYNAMO, GPSS II, SIMULA and SOL, which can be used as per their performance and availability to the users.

DATA GENERATION AND DISTRIBUTION

As discussed in the above paragraphs that to get authentic and accurate information regarding maintenance related functions is really difficult if not impossible. Therefore, a need for generating random statistical deviates from certain distribution is felt. The distribution will be the one that adequately describes and represents the physical process involved at that point in the experiment. During an actual simulation experiment, this process of generating a random deviate from a particular distribution may have to be done many times for many distributions, depending on the complexity of the model being investigated by the experiment.

The general process for generating a random deviate from a specific distribution will nearly always follow, this pattern:

- Generate a random number from the uniform distribution.
- Perform mathematical transformation of the uniform random number or numbers which produce a random deviate from the described distribution.
- Use the transformed deviate in the experiment as required.

These three steps are then repeated many times for the chosen distribution as the experiment proceeds in time. After the numbers have been generated using any of the three methods (Mid product, Febonacci and Mid Square), the same are to be tested by Chi-square test and the Kolmogorov, Smirnov test. To save time during testing one of the methods (the gap test, the runs test, the Poker Test) can also be used.

Through an investigation of the simulation output, it is possible to create a response surface in which the various factors involved are designated at different levels of operation. In order to characterize these surfaces experimental designs are often used to analyse, quantify and predict the effects of response surface change. Experimental designs such as full factual, Latin squares, fractional designs and rotateable design can be used for digital simulation. The primary purpose of the experimental designs to determine which variables, are most important, how these variables influence the response of simulation model and why certain results occur as they do in the simulation experiment?

Variance Reduction Techniques

Since the simulation itself is a random process, the output is by definition is a random variable, which contains statistical fluctuation and great care is needed to use these results. A variance reduction technique is therefore used, which reduces the inherent random deviation in our statistical output. Some of the techniques used include, use of expected values, stratified sampling, importance sampling, control variates, antithetic variates and quasi random numbers.

Statistical Analysis of Simulation Output

The statistical analysis is needed to validate, authenticate and collate time series data, which are generated by digital simulation models. Validation means real world situation, authentication denotes the establishment of a measure of confidence in a single set of results and collation connotes a critical comparison of two or more set of model output.

To describe a random data four statistical function namely, mean-square values, probability density functions, auto-correlation function and power spectral density functions are required. The purpose of a statistical analysis of simulation output would be to remove the effects of autocorrelation from the data and in the process detect other abnormalities.

Optimization of Simulation Parameters

Optimization of simulation parameter is necessary to know the effect of other co-related parameters. For example, in maintenance system it is not sufficient to only find out the size of maintenance crew but also its total cost. Direct search techniques, which require only a black box response, can be used to guide a simulation to an optimum response with respect to chosen decision variables.

SIMULATION APPLICATIONS

Simulation technique can be applied to many of the maintenance related functions, which include inventory of spare parts, manpower estimation and so forth.

Simulation of Inventory

The factors related to spare parts inventory for maintenance function include, economic parameters, demand of spare parts, their ordering cycle, lead time, planning period, number of suppliers, stages of inventory, number of items with their availability and the Govt. Policy. All above factors are responsible for the development of a good inventory system. In maintenance, it will be really difficult to get all information as per requirements. There are various inventory models available for conventional systems when their nature is deterministic as well as probabilistic. However, it is easy to maintain stock for planned preventive maintenance since the requirements for each and every part/component are known in advance. But in case of breakdown maintenance everything is uncertain since failures are random in nature. Therefore, knowing the spare parts requirement is difficult and simulation can be used to predict the number of failures and inventory requirements based on the past information.

When failure time and repair time of the equipment are known, simulation can be used to know the completion time of the jobs within a specified time period i.e. simulation time. The waiting time of jobs or mechanics can also be known to make alternative arrangements. Policy decision regarding the type of maintenance system to be adopted under particular situation can also be known with the help of simulation.

Manpower Planning

Maintenance functions are most unpredictable in nature and to know their exact time period is highly flexible and uncertain. The estimation of manpower requirements for such jobs becomes very difficult. When planned preventive maintenance system is being used, calculations are slightly easy but when random failures are to be taken into account manpower planning is a complex problem. With the help of computers simulation models can be developed to take care of such eventualities. Though the results obtained will be sub-optimal but compromise can be made with the situation. For this following steps will be required:

- Calculation of failure frequency equipment-wise.
- Number of equipment due for periodic maintenance month/week wise.
- Skill of work force required i.e. skilled semi-skilled and helper, etc.
- Scheduling of maintenance work.
- Calculation of manpower requirements.
- Spill over/non-completion of work during specified time period.
- Cost optimization i.e. to estimate optimum number of crew size.

Following the above steps a computer program can be run for different set of combinations, considering change in number of equipment and percentage variations in spillover for a specified time period i.e. six months or one year. The results obtained will give fairly a good idea regarding manpower requirements day-wise, week-wise as well as month-wise. This will help in providing optimum number of maintenance crew so that the maintenance costs can be reduced. Looking in to the need of particular equipment, practices for maintenance work can also be set without distributing the basic schedules.

Through simulation, an investigation can be made to establish suitable performance indices to represent the functional behaviour of the system under operating conditions. The conclusion drawn can be useful to take precautionary measures necessary for minimization of downtime of the equipment/system.

It is practically noticed from the above discussions that the use of simulation helps indirectly for the assessment of reliability of the system/equipment. Simulation helps in acquiring relevant information of failure of components/parts used in equipment. Having obtained the desired information the same can be used to make reliability calculations. Before starting the simulation process the nature of the failure distribution must be examined to generate the required data. If the information's procured do not satisfy the constraints of the working environment the results obtained will fail the real life situation.

Reliability evaluation is purely based on the failure rates and the time period for the system/equipment will be available uninterrupted. Therefore, the authenticity of the generated information is most important to yield realistic results. With help of simulation studies programs can be drawn to schedule the maintenance functions at appropriate time intervals to avoid premature failures. Attempts should be made to generate and test the available data for accurate reliability predictions, otherwise the efforts will not yield the desired outcomes.

QUESTIONS

1. Discuss how simulation can be used for reliability analysis.
2. What type of problems are faced during reliability evaluation?
3. Describe simulation procedure for reliability calculations.
4. How data should be collected for effective simulation process?
5. Discuss how simulation model are designed.
6. Highlight the importance of simulation in maintenance systems.

8

Reliability Improvement

INTRODUCTION

The use of automated production system demands for high reliability of the equipment/system. However, before purchase of a new equipment/facility, the following points should be given due considerations to improve the reliability of the system. For any new organization it starts from the very basic i.e. procurement procedure, specification of equipment, vendor selection and bid conditioning machinery reliability audits and review etc. The use of maintenance also plays an important role for the improvement of equipment/facility reliability. The other methods followed for the improvement of reliability include, load conditions, attitude of material supplier, spare parts and their effect on service life of equipment, modern maintenance techniques, maintenance cost reduction and role of lubricants on reliability. In some cases, special seals for pumps and motors can improve reliability and can also provide better safety for the system.

As it is seen in practice that no system can work failure free, irrespective of any type of maintenance being followed. Undoubtedly, the type of maintenance system can improve the reliability of the equipment to some extent. However, if an equipment/system has to work uninterrupted, the concept of redundant and standby system has to be incorporated with the system. For example, if main power supply from the source fails, a generator power supply has to be operated. In most of the power generation units, the concept of standby unit is being practiced. Though it is a costly proposition but has to be adopted to meet out the emergency requirements and also to improve the overall reliability of the system. Now we will discuss above cited points one by one.

SPECIFICATION REQUIREMENTS

The job, to specify the machinery should not be taken as a routine matter to either give superficial information or to give bulky specifications, which may burden the supplier. A good specification is therefore, concise and precise. It should define requirements in clear and understandable form so that it encourages the vendor to offer more than the bare minimum requirements. Resourceful and forward looking equipment purchasers or owner companies should require the manufacturer of vendor to develop and provide machinery installation instructions, equipment commissioning instructions, proposed instrument checkout guidelines or sequences, equipment start up instructions and other helpful documents. If these requirements are included at the initial stage, will cost less as compared to the later stage.

The most effective method of specifying equipment is the single narrative document in place of disjointed individual specifications. The single narrative specification should try to uncover oversights, weakness and deficiencies in the procurement and design efforts for the equipment. With the bulky specifications it is difficult for the vendor to find relevant specification clauses and other related information. The most effective way of reducing failure risk at the specification stage is to examine vendor experience. Detailed experience list should be submitted for the purchaser's review. These lists must clearly identify the locations where similar type of equipment have been supplied and installed. A report from the said users can be obtained to verify the claims of the suppliers.

Inquiry documents and purchase orders should make it clear that full compliance with the specification will be contractually binding requirements, unless the vendor has taken specific written exception to a given clause. The assessment of first cost verses maintenance cost should be made to keep the equipment/system running at low maintenance cost in future.

Even the best specifications may be lost on a vendor who lacks skill, experience, and resources and follow up organization. For the vendor to carry out the manifold detailed engineering, manufacturing and quality control steps that will culminate in high quality product, a comprehensive statement on specification is a must.

The enquiry specifications should be sent to such vendors only who can demonstrate, compliance with the specified requirements and can stand the test of rigorous bid comparison with his competitors. In general, the machinery vendors should excel through past satisfactory experience in design, execution and after sale service of their products.

SYSTEM RELIABILITY AUDITS AND REVIEW

The basic need of equipment/system audit arises to check the problems related to fundamental design of equipment. The other objective could be, to subject the design, to non-routine computer analysis to test its validity whereas, the reliability reviews are aimed at ensuring compliance with all applicable specifications. These reviews also examine for acceptable deviations from defined specifications. Reliable and efficient equipment is probably most important factor for ensuring profit in any production organization.

It is always advisable to ensure reliability aspects of the equipment before its delivery to the organization, but it is really a difficult task. Since at a later stage it is difficult to incorporate any modifications to improve the equipment reliability. It is observed through experience that it is possible to check the reliability of the equipment provided some failure statistics is made available.

An experienced audit team should be careful to keep the focus on existing as well as potential reliability oriented observations and recommendations. The equipment observations, which are cost oriented, should be cared for. For example, periodic lube oil replacement vs periodic on stream purification of turbo machinery. This will become an issue if the lube oil is not replaced at the required interval. The equipment audits provide state-of-the-art, updates, technology transfer and greatly reduced risk of unexpected plant shutdowns due to equipment/system failures.

Through reliability test, which are to be carried out under actual environmental conditions such as temperature loading condition, etc. can justify the claims made by the manufacturer/supplier. The performance of equipment/system can be verified at all levels as per the recommendations made in the purchase order. These tests can also reveal pre-mature failures of the equipment/system. For example, the performance of radiator used for cooling internal combustion engines will largely depend upon the environmental conditions i.e. temperature.

IMPACT OF MAINTENANCE ON RELIABILITY

The reliability of equipment/system can be improved through better maintenance practices. But the cost of such maintenance related activities must be compared with enhanced availability of the equipment. Through studies, it is shown that with increase in maintenance costs, the mechanical reliability of the equipment improves to an appreciable level, but this can also be achieved by better performance and management approach. The various practices and policies associated with low cost high reliability performance include the following:

Organizational Goals

It is the main purpose of any organization to minimize downtime of equipment; it is bound to improve the availability of the system. Consequently, the organizations are to be designed to achieve the predefined goals. It is observed that the ease of managing the maintenance activities should, therefore, be given lower importance. If the skilled operators can be trained on maintenance aspects also will definitely reduce the downtime of the equipment/system.

Information System

The physical needs of any system can be described through better information system. This includes, failure history, man hours spent on maintenance costs and operating performance of the equipment, which must be cared in time. The above-cited information can be used for making in-time decisions, which may be useful in enhancing the life of the equipment as well as their operating performances, which ultimately improves the overall equipment reliability.

Technology Up-date

Though there are many advance techniques available for better maintenance, which include, reliability centered maintenance, sneak analysis etc. that can be used to improve the equipment reliability. But many of the above techniques are not being practiced in actual working environment because of limited manpower, time and money. Therefore, the new techniques must be reviewed critically with special interest and be applied in all functional areas to get the real advantage out of them.

Review of Reliability Activities

Whenever, some reliability improvement techniques are being used in any organization, it must apply to the whole system and not in isolation to some particular area only. The actual benefits can only be achieved through holistic approach of applying reliability improvement techniques because it is important to improve the reliability of the total system. However, emphasis must be laid on the critical equipment or component depending upon its importance in the organization.

ACHIEVEMENT OF RELIABILITY

In the process of improving the reliability of equipment, the concept of maintenance be based on the reliability-centered maintenance. The culture of an organization is the set of values, behaviour and standards, which make it to success in present environment. Reliability can only be maintained/developed if the management are committed i.e. are willing to install an appropriate reliability management structure, are willing to accept responsibility for reliability and are willing to improve the necessary resources. Everybody in the organization has to be reliability conscious and is willing to accept changes in working methods and is willing to accept appropriate attitudes. The three essential elements of methods, management and workforce are discussed in more detail.

Reliability Methods

The life cycle for an engineering product passes through different states, starting from specification, design, production, distribution, installation, operation and maintenance, to final disposal after use. Under this life cycle for a 'Reliability System', there are set procedures, applicable to each stage to improve the reliability of the final product. The following are some important examples of these procedures:

(*a*) As stated earlier, the product must be manufactured as per the required specifications.
(*b*) The design should be such so that it meets the environmental hazards during all working conditions.
(*c*) The testing of the design, to ensure that it meets the specifications as intended.
(*d*) Specifications of all components and materials to be procured must be verified.
(*e*) Proper selection of the suppliers for components and materials must be made.
(*f*) Verification of in-coming materials and components for meeting out documented specifications.
(*g*) Verification of production process to meet out design requirements.
(*h*) Methods to ensure that manufactured equipment is set up and maintained properly to perform it's desired functions.
(*i*) Inspection and testing of product in process as well as after completion to ensure its quality.
(*j*) Full analysis of the root causes of faults and correlation to prevent any re-occurrences.
(*k*) Acquisition of data on the operating performance to check, verify and reliability standards.
(*l*) Use of all feedback data, in the development of the specification for a replacement of product in future.

MANAGEMENT OF RELIABILITY

The management of reliability must be directly under the control of maintenance engineer/manager. The maintenance engineer must ensure the following for reliability improvement:

(*a*) The setting of reliability objectives, which must match with the objectives of the organisation.
(*b*) The formulation of a reliability system to ensure the availability of the equipment for effective use.
(*c*) The implementation of reliability techniques, to ensure that the set standards are finally achieved.
(*d*) The preparation of reliability manual, which should define reliability system in practice.
(*e*) The education and training of men who are involved in maintaining the equipment and reliability.

(*f*) The documentation of relevant data for preparation of maintenance schedules for ready reference.
(*g*) Maintenance of adequate records covering all details to be used in future to either carryout planned maintenance or to improve reliability.
(*h*) The taking of corrective actions in all situations, which could affect the equipment reliability.
(*i*) The auditing and review of reliability system time to time for its effectiveness.
(*j*) Identifying and monitoring the costs benefit analysis to support the reliability system.

WORKFORCE COMMITMENT TO RELIABILITY

The first responsibility of the management is to clearly understand the objectives and methods of reliability systems and to examine that workforce under their control is capable of implementing them. The next important and difficult problem is to get the workforce to accept full responsibility for ensuring high reliability of their products. This involves motivation to set standards to achieve high reliability with due criticism from fellow members or outside persons.

Here the role of the maintenance engineer/manager is very important since he has to act as interpreter who can translate the output of applying knowledge to work into daily activities that are to be performed by the staff under his control. The responsibility of maintenance engineer is to set the goals for maintenance related activities and monitor to achieve them. The various functions, which can be undertaken by an able maintenance engineer, include the following:

- Failure mode and effect analysis,
- Fault tree analysis,
- Distribution analysis,
- Interpretation of equipment availability modeling,
- Management of preventive maintenance program,
- Cost benefit analysis,
- Development of new maintenance strategies,
- Failure analysis,
- Maintenance task analysis.

The information available with the maintenance engineer can be utilized to control the equipment/system availability in a real time mode. The knowledge and the expertise of maintenance engineer will be useful for the success of any modern organization.

LUBRICATION AND RELIABILITY

Equipment/plants where pressure lubrication is being used, the purification of lube oil is very important, as a huge quantity of oil will be required. It is observed through experience that self-contained lube oil reservoirs are subjected to contamination and deterioration from particulate matter and water. The gas compressor lube oil systems are further exposed to potential dilution from higher hydrocarbons, which change the viscosity and flash point of the lube oils. For purification of lube oil, dewatering centrifuges are used. Today, the use of coalesces is very common but it needs more maintenance.

The use of synthetic lubricants is gaining momentum and it is believed that with their improved properties, it is also economically viable. This is possible only through additives, since they change many properties of the oil and it can be re-used easily. Therefore, the equipment/system failures are minimized and therefore improve its reliability.

RELIABILITY AND MAINTENANCE MANAGEMENT

The maintenance cost as a percentage of the estimated replacement cost (ERV) is about 4.60% but in practice it should be 3% of ERV whereas in process industry it is 4.2%. Through comparison it was noticed that the maintenance cost for the similar type of products for the same market being produced by two different companies vary significantly. The reason for this variation is the guidelines for the maintenance being followed.

It is difficult to establish a reliable estimated replacement cost (ERV). This will depend on the initial cost of equipment and the index followed for current rate of replacement value with depreciation cost per annum. Efforts are to be made to improve competitiveness of productivity. So, whatever has been measured and compared has to be put in actual action plan "with execution to improve the number used for comparison.

The maintenance practices followed are old but should be done in efficient manner to deliver better performance. In industries the same have been used to evaluate and identify maintenance improvement opportunities effectively. But now it is important to describe that what are the best practices, the execution or how can vary between plants and also with-in the plants. It can help to understand the complete maintenance process to uniformly evaluate essential reliability and maintenance process.

The key processes include leadership, support facilities, preventive maintenance, planning and scheduling, material management, technical data base skills development etc. To improve the reliability of a system the following points must be considered under planning and scheduling of maintenance:

1. The work request with ID number is to be submitted in Computerized maintenance management system (CMMS).
2. Prioritization according to set guidelines.
3. Equipment number, originator, description etc. to be listed.
4. A well defined work description or observed symptoms recorded.
5. Preliminary work order type should be defined as either corrective maintenance or system improvement.
6. How the work affects the process, is identified according to the recognized maintenance opportunities.
7. Work request is validated.
8. Work flow process is followed.
9. Known safety requirements are identified.
10. Work request can be entered.
11. Work requests are filtered and collected at a central point.

To examine any maintenance system the aforementioned elements must be used to evaluate own practice in the initial part of planning and scheduling.

With any maintenance program and improvement initiatives the role of proper planning and scheduling is very important and can play a key factor in cost effective maintenance practices. Even the programs like reliability centered maintenance (RCM), total productive maintenance (TPM) reliability based maintenance (RBM) can support the importance of planning and scheduling provided the basics of it are understood by the users. The first stage of planning includes final scope of work, safety requirements, major steps to work, important clearances, spare parts needed and secured as available for when the job is scheduled, specified tools, skills required, time needed to do the job and so forth. Secondly, scheduling means to decide when the job will be done and who will do it.

The following are the effective guidelines for planning and scheduling:

(*i*) Planning is done before scheduling
(*ii*) Planning and scheduling are done before execution of work
(*iii*) Scheduling is done for the work to be done
(*iv*) When executing a planned and scheduled job, people are not interrupted to do other work
(*v*) A job is not finished before documentation, why the job had to be done
(*vi*) Find out the root cause of any identified problems

If the above planning and scheduling practices are implemented will result in less use of outside contractors, less unscheduled overtime, increased overall equipment efficiency (OEE), less unscheduled downtime and more free time to perform root cause failure analysis.

The other factors used to improve the reliability of system through maintenance are the following:

Correct prioritization of work: This must be done carefully and correctly with the consequences of not doing the work before the given time. In a plant with multiple product lines it is necessary to know the most important line for the desired production to meet out the customer demand on time. Disciplined priorities will also lead to correctly performing planning and scheduling tasks. Check that the contents of the Preventive Maintenance/Essential Care and Condition Monitoring (PM/ECCM) are Night and the consequences of a failure must be more "expensive" than the cost of trying to prevent it. The right contents also include using right methods for basic inspections and condition monitoring. Most of the PM/ECCM should be done while the equipment/system is operative and very little should be done while the equipment is idle. The frequency of PM/ECCM be based on failure data statistics to make it more cost effective.

APPROACHES TO IMPROVE RELIABILITY

In a general approach improvement is widely applicable to reliability improvement. The main area to improve reliability is the design phase, where reliability engineer can help defining areas to improve reliability through alternatives available to him. However, the following action plans can help in this regard.

(*i*) The user's needs be reviewed critically to minimize the unreliable parts/components from the system. It is known that the availability of equipment is directly associated with reliability which can be enhanced by better and timely maintenance practices.

(*ii*) Use of standby systems can minimize total down time of the equipment.
(*iii*) Selection of standard quality parts/components can improve availability of system and therefore the better reliability of the system.
(*iv*) Avoid over loading/over stressing of parts/components in use to enhance their useful life period and ultimately reduction in failures.
(*v*) Control the working environment which might affect the equipment life and therefore reliability.
(*vi*) Properly monitor the maintenance practices being followed to repair/replace the defective parts in time which may lead to further deterioration of the system.
(*vii*) Use robust design methods for the products being deployed under terrain conditions.
(*viii*) Screening tests should be carried out to detect infant mortality failures and to eliminate sub-standard parts/components.
(*ix*) Conduct research and development to attain improvements in the basic reliability of those parts which are most unreliable.

The above cited points don't give authentic solution to improve the overall reliability of the product but it definitely helps. The coordination and cooperation of production and operating personnel will certainly be of immense help to improve the reliability.

QUESTIONS

1. Suggest the methods required to improve the reliability of systems.
2. What is the importance of specification requirements in improving reliability?
3. Discuss how maintenance affects the reliability improvements.
4. How desired reliability can be achieved effectively? Discuss.
5. What is the role of maintenance managers in improving reliability of components/systems?

9

Reliability Assessment

INTRODUCTION

The systems comprise of number of components placed either in series or parallel for performing a specific task without interruptions for a stipulated time. For this purpose maintenance activities are to be accelerated to avoid unforeseen breakdowns. The reliability assessment here means to know the time period for which the system will work without failures/ breakdowns. By carrying out case studies on the past performances of the equipment, it is possible to predict its failure free time periods. For this purpose, certain past data will be required for the equipment under investigation. The authenticity and accuracy of the above information will yield correct results, which will be close to real life situation. There are a number of methods available to assess the performance of equipment/system for intended time period. The first information required will be the failure data of the components/units of an equipment/system.

RELIABILITY PREDICTION

The reliability requirements suggest that the estimate or prediction of reliability before the actual production of equipment/system should be carried out. Prediction of reliability is a continuous process to be carried out at various levels of design starting from the design of product until it is put in use. Based on the past history of equipment it is easy to predict the reliability which is basically dependent on failure rate distribution and the working conditions of the product. It can be on Exponential and Weibull distributions since failures follow these distributions very closely for most of products.

The following steps can be used to predict the reliability of the equipment/systems in use.

(*i*) **Definition of system:** Any equipment/product is made up of sub-systems/components, units which must be specified for their functional configurations and boundaries. This information will help to make the functional block diagrams of the products/equipment to assess the reliability. The failure of components in a block will make system failure depending upon its configuration.

(*ii*) **Block diagram:** It is similar to functional block diagram with the difference, to identification of reliability influencing factors. This shows the sequence of operations of components with their importance on reliability parameters. A complex system which is made up of thousands of components, however, units which are not essential to successful operation need not be included such as decorative accessories as their failure do not affect system reliability. Redundant paths and alternative parts/components are clearly specified in the diagram.

(*iii*) **Identification of relevant factors to reliability:** Here emphasis should be laid on critical parts/components which are crucial for reliability prediction. Factors which are responsible for the failure of parts should be examined carefully and correctly. Adjustment of parts/components may play an important role to improve reliability which must be cared for.

(*iv*) **Selection of reliability data:** The failure rate data of all critical parts with respect to time and environmental conditions must be examined properly. Collection of such data is difficult task at initial stage particularly for a new product. However, this information can be collected from field performance studies, life test, inspection tests and any other sources.

(*v*) **Determination of reliability:** Using above information the reliability of each block component/part can be calculated, which can be used to determine the overall reliability of the product/system. In case of non-availability of relevant reliability data accurate estimates based on past experiences of similar components can be made. In such cases simulation models can also be applied.

(*vi*) **Application of predicted results:** This helps to predict the successful life of product but also indicates the weak areas, the extent of improvements required and alternative methods which can be used to improve the reliability functions.

THE RELIABILITY DATA

The starting point in reliability assessment is the evaluation or estimation of reliability of component or unit that can only be done with the help of failure data of the component. The failure of a component means that at a particular instant it is not performing its assigned task. Under certain conditions the items/components can be repaired and reused, whereas some components can only be replaced, as no repair is possible. The inaccuracy in reliability data may not yield the correct results even after using the correct model. This inaccuracy may arise in obtaining the data or in their use. The availability of data in the required form is therefore essential otherwise their synthesis will be required.

The reliability data can be obtained from field experience and sample testing. Field experience is meant for the monitoring and recording of the elements reliability characteristics over a period of practical usage. Sample testing on the other hand, refers to an experiment deliberately set up to record the required information from a representative sample. The obtained failure data must be tested within the specified confidence limits to make the generalized representation, since the data pertain to a specific component unit. This will give an indication about the distribution of the failure data, which have to be collected.

RELIABILITY MODELS APPLICATION

Depending upon the configuration of parts/components different reliability models can be used to evaluate the reliability of equipment/system. A particular model can suit only a particular layout of components in an equipment or system. Based on the layout of the components, the reliability of the equipment/system can attain different values. In case of series layout of parts, the system will fail by failure of any of its component/part and its reliability will fall to zero whereas, in case of parallel layout, system may remain active even after failure of its some of the components/parts. But from design point of view, it is always not possible to select any particular layout of components/parts. However, efforts can be made during design phase to take care of component layout also, which will directly affect, the reliability of the equipment.

Though various models for reliability evaluation are available, but its application will largely depend upon the configuration of the components. In case of series model, the reliability of each components/parts will play an important role since equipment reliability will be the product of the reliabilities of each component. However, in case of parallel configuration, it will be the summation of reliability of each component/part.

It is noticed that when redundant components/parts are used with any equipment its reliability improves significantly though the cost of such systems would be high.

RELIABILITY OF MECHANICAL SYSTEMS

The causes of mechanical equipment/part failure can be many, which include load, force, fatigue, mechanical wear, interference fits, bolted joints, antifriction bearings, welded joints, etc. When force or load is applied on the part, it is stressed and this stress is the cause of failure of the component/part. The strength of the part depends on variety of factors, including design, material properties, material imperfection, fabrication process, shape and size. The reliability of such component can be expressed in terms of stress and strength. The stress in the system may be due to direct axial or shear forces, bending moments or torsion, etc. Similarly, the strength may be normal or shear strength, flexural or torsional strength, etc. The reliability expression for such system can be expressed as:

$$\text{reliability} = R = p(s > w) \quad \text{where strength } s \text{ is greater than load } (w)$$

$$= p(s - w > 0)$$

The reliability of the component can also be computed in terms of stress and strength.

Depending upon the operating conditions, the mechanical components/parts are subjected to different kinds of forces/loads, which may be static or dynamic. The shape, size and hardness of the material will have different forces on operating machines. The terrain conditions for vehicles will impose variety of forces on the machine elements. Dynamic loads occurring in toothed gears and other transmission devices are very common. In engine operations, sudden braking may cause impact loads on the machine components. The total load spectra, though very diverse, can still be classified into a finite number of categories for the evaluation of reliability. The loading pattern can have various shapes depending upon the conditions. From these shapes the type of distribution, a loading pattern, can be known to apply in reliability models. It is seen from experience that light, extra light and heavy force conditions have Beta distribution, medium force conditions have either uniform or normal distribution pattern.

When a machine element is exposed to fluctuating or variable loads, the possibility of failure due to fatigue should be examined. These can be examined in the form of little permanent deformation, brake marks showing progression of the crack seen by microscope, and the brake marks, which are smooth as a result of rubbing to assess the possibility of failure. The example given below will highlight the application of reliability model of a mechanical system. The use of dumper is very common in the open cast mines for shifting the material i.e. coal or overburden. For the purpose of study, the dumper is divided into following three functional units:

- The proper generation unit (DPGU) comprising of diesel engine.
- The dumper power development unit (DPDU) comprising of compressor and hydraulically controlled valve bank and
- Dumper power utilization unit (DPUU) comprising of direct mechanical torque converter system for propel operation, hydraulic cylinders for steering/hoisting and hydraulic cylinders for dumping operations. The block diagram of dumper reliability model is given below:

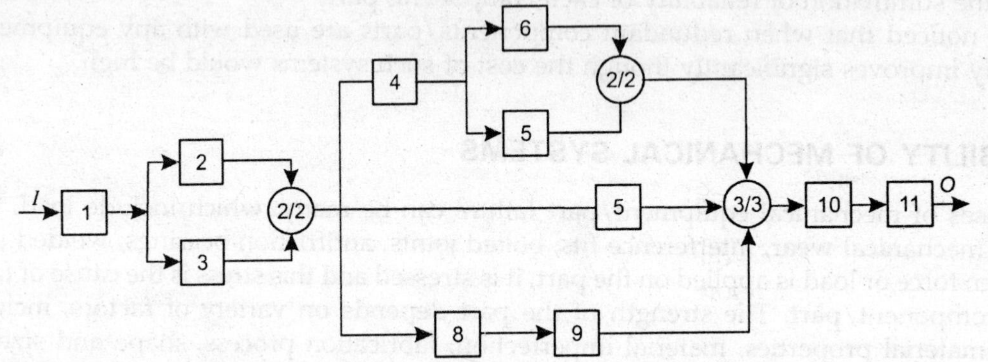

Fig. 9.1: Dumper Reliability Model

1 – Engine;	5 – Brake;	9 – Suspension cylinder;
2 – Compressor;	6 – Clutch;	10 – Body with canopy and frame;
3 – Valve Bank;	7 – Steering cylinder;	11 – Wheels/tyres
4 – Transmission;	8 – Body left cylinder;	
	I – Input	O – Output

The expression for dumper reliability (R_D) is obtained as:

$$R_D = \text{RDPGU. RDPDU. RDPUU} \quad ...(9.1)$$

Where, RDPGU – Reliability of power generation unit
RDPDU – Reliability of power development unit
RDPUU – Reliability of power utilization unit

In the similar manner, the reliability of any complex mechanical system can be evaluated evolving any number of component/equipment.

RELIABILITY OF ELECTRICAL SYSTEMS

To maintain uninterrupted power supply in any system is essential for better production/ service. Therefore, reliability of each component used in electrical system is vital and significant. Most of the electrical systems work in two states: (*i*) successful (working), and (*ii*) unsuccessful (failed). If no other variations exist, then the system reliability is simply the probability of the system being in the successful state and can be expressed as:

$$R = p$$

and in the other state the probability of the system is zero as it is in the failed mode.

The causes of failure of integrated circuits (IC), can be visualized by the diagram as given below:

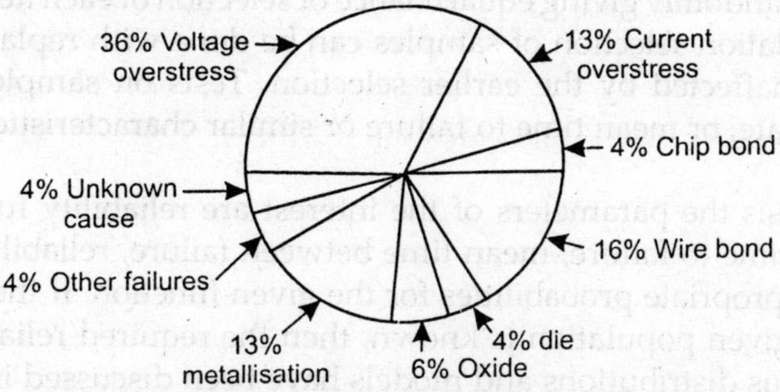

Fig. 9.2: Failures of Integrated Circuits

It is seen from the above diagram that maximum failures are due to electrical stresses and amount to about 49%, therefore, due attention should be given for this mode of failure.

ELECTRICAL STRESS

In addition to temperature, some electrical quantities such as voltage, current or power effects are also the part of increased failure rate. In order to lengthen the useful life of the operation of parts is recommended not with maximum acceptable ratings but with reduced stress conditions. Such rating will improve the reliability and minimize the failure rate.

It is seen that voltage factor increases with rise in voltage. The failure rate of some of the IC's is proportional to the factor. With power supply of 15V a, CMOS IC may fail 10 times as much as if it is operated at 5V which clearly shows that how variation voltage affects the system performance.

A serious problem is the electrostatic discharge (ESD) damage which is termed as the new contaminant of the age of microelectronics. It occurs in both MOS and bipolar devices. It can cause an immediate or delayed damage to circuits. In many cases no damage can be measured until months have passed and the equipment has failed in the field. A voltage, as low as 100V, is reported to be capable of damaging semiconductor devices. Overstress (Fig. 9.2) is assessed to have occurred by ESD. This happens in spite of input protective devices (e.g. diodes) on the IC to prevent the discharge from supporting the gate oxide. The other provisions to protect against ESD are:

- a proper floor finish,
- grounding of furniture,
- application of special sprays and
- control of humidity in the work areas.

The other kind of damage is caused by current stress. The main effect is electro migration of metallization patterns. The rate of migration is proportional to current density and failure rate increases approximately with the square of current density.

RELIABILITY PARAMETER ESTIMATION

If the failure distribution of relevant associated parameters are known then the parameters required for the reliability can be obtained. When the population of the desired data is large sampling can be done randomly giving equal chance of selection of each item in the population. For finite size of population selection of samples can be done with replacement so that each selection process is unaffected by the earlier selection. Tests on samples are conducted to determine the failure rate, or mean time to failure or similar characteristics appropriate to the population.

In reliability analysis the parameters of the interest are reliability function, hazard rate, failure density, mean time to failure, mean time between failure, reliability life etc. All these are associated with appropriate probabilities for the given function. If the failure distribution function governing a given population is known, then the required reliability characteristics can be obtained. Various distributions and models have been discussed in previous chapters. The two important quantities generally obtained from any distribution function are mean and variance, which can be expressed in terms of parameters for any distribution.

RELIABILITY SYNTHESIS

For the evaluation of system reliability collection of appropriate reliability parameters is essential. But the collection of relevant information is difficult at all levels. If the data are not available at the system level, but are available at a lower level, through systems synthesis, the same can be used to complete the system. The reliability model, which combines both discrete change of state and continuous variations in performance, needs to be examined for methods of synthesis. Here for the purpose of simplicity, the two-element system can be considered which will have four possible states. In general, if the system can be represented by just two states than sub-states act of the four will combine for success and the remaining (4-n) will combine for failure. Since the two-element system can take on a number of forms depending upon the logical arrangements, it is required to consider them separately. Where both elements are required for system success, can be termed as 'non-redundant arrangement'. When only one element is required for the system, system will be called as redundant. The non-redundant system can be considered as series reliability model, whereas redundant system will fall in the category of parallel reliability model.

In the same manner, the complex systems can be reduced to standard form during the reliability parameters of interest.

RELIABILITY IN WELDED JOINTS

The use of welding is very common for joining two metal parts permanently in the present industrial environment. Therefore reliable welds are required for every weldment produced. Many welding techniques are being practiced to assure weld quality or reliability. Each of the welding methods provide guidelines to obtain reliable welds after proper testing. The main factor influencing the quality of the weld is the skill of the welder when manual welding is being performed. The purpose of a welding procedure qualification is to show that the proposed weld joint of weldment will have the required properties for its intended application. The document should contain the actual welding variables used to produce the test weld and the results of the test conducted.

The purpose of the welder performance qualification is to determine the ability of the welding operator to produce sound weld metal following a welding technique as per specifications. Quality welding on a product must be judged with respect to a specific quality standard that is based on its intended service. It must be a balance between the service requirements and the consequences of failure versus economic factors. Normally the weld quality requirements are controlled by codes, specifications, and standards, however in absence of it the welder must make the products of high quality standard in order to meet the market challenges. The cost of weldment is also an important factor.

The major problem faced in welding is the doubt of the design engineer, that the product is not fabricated as designed. This can be taken care by the quality assurance program, which saves the money in the long run as premature failures are avoided.

FAILURE ANALYSIS IN WELDING

It is experienced that the failure of large welded structure is rare. When failure of such systems occur investigations are carried out to avoid repetitions of similar problems. It is important to make objective study of failed parts/structures to know the cause. This is done by investigating the service life of equipment/system, that led up to the failure and the actual mode of failure. Investigation should be carried out to reveal the cause of failure. Which may suggest likely changes in design, manufacturing or operating conditions to avoid similar failures in future. Sometimes the suggested changes may assure a more reliable product. The following areas must be examined to determine the cause of failure and other factor involved.

(*i*) **Primarily observation:** the study team should make thorough investigations of the actual part/component which failed at the working site at an earliest, after the failure has taken place. This will help to examine the failure surface, fracture texture, appearance etc. in detail. The persons present at the site should be cross-examined to know the details of failures.

(*ii*) **Background information:** The study team should collect all the relevant information regarding specifications drawing, component design, fabrication methods, welding procedures, weld schedules and repairs if carried out during the service life. The working conditions of equipment which has failed must also be examined critically which may include operating temperature, service loads, overloads cyclic loading abuse and so on.

(*iii*) **Laboratory studies:** The study team should see that laboratory tests are carried out on the failed parts to verify the material composition, mechanical properties etc. The failed parts should be investigated properly to collect information which may be useful for design/ modification and fabrication of the parts. Other defects in the parts noticed should be examined though they might not have contributed towards the said failure.

(*iv*) **Failure assumption:** The study team should not only look into the positive facts and evidence of failure but should also see, all the negative responses that may be learned about the failure. The decisions should not be arrived at with certain specific assumptions only. The actual failure should the synthesized to include all available evidence, this might lead to the need for collecting additional information about the failure.

The true cause of the failure will be known from the investigation carried out. The assumptions considered must also be examined to assure the actual cause of failure which could be one out of the following:

(*i*) Faulty design failure or incorrect selection of material
(*ii*) Manufacturing fault due to men or machine
(*iii*) Failure due to wear out

The failure due to faulty design or incorrect selection of material involves failure due to inadequate stress analysis or calculation errors during design for the propose of loading etc. The failure must have been caused due to improper weld joint.

Failures can be due to incorrect processing or poor workmanship. The quality of the weld may be sub-standard. Failures can also be caused due to poor fabrication practice such as incorrect filler material used in welding.

The overloading is one of the major problem to cause a premature failure. Corrosion due to environmental metal conditions can also add to stress concentration and ultimate failure. Poor maintenance could be another cause of failures. Environmental conditions do play an important role which must have not been considered during the design stage.

QUESTIONS

1. Discuss the need of reliability assessment.
2. How different reliability models can be applied in real life situations? Discus briefly.
3. With example discuss the reliability of an mechanical system.
4. How reliability of an electrical system is different than mechanical system? Explain in detail.
5. Discuss the reliability synthesis and highlight its importance.

10

Reliability Testing

INTRODUCTION

To meet out the design and reliability specifications it is desired to test the finished product so that the failure rate remains within the pre-specified lower and upper limits. The problem of testing is more significant where the life time of the component is very long normally in case of electronic items. In case of such components, the under actual usage condition would require too much test time and excessively large sample size. Accelerated test are therefore used for the primary reliability-related factors of applied voltage, temperature and humidity. In addition, statistical sampling is used, taking into account the similarities between products in process and design, so as to minimize the number of test samples.

Reliability testing should be considered as a part of integrated test programs which include the following:

1. Statistical testing: to optimize the design of the product and the manufacturing processes.
2. Functional testing: it is to confirm that the design meets the performance requirements.
3. Environmental testing: to ensure that the designed product can work successfully under specified conditions.
4. Reliability testing: to ensure that the product renders services for the specified period of time without undesirable problems.
5. Safety testing: to ensure all safety aspects have taken care off.

Reliability is a function of time and variability therefore, the tests should be carried out for a long period of time, to ensure failure distributions, particularly wear out failures. It is observed that if the tests are carried out properly it will bring down the total development cost and reduction in overall maintenance functions.

RELIABILITY AUDIT

Before any product is put into the market, it must be tested over a period of time to measure its achieved reliability with due considerations of the environmental conditions. But these methods are not being practiced by many of the organizations particularly in India and therefore, their products don't compete at international market. The need for reliability audit is more pertinent under such conditions to promote the sale of the product/equipment. For every product national/international standards are set by the quality control departments and the same should be examined through internal or external audit teams. Even internal audit is a good practice if followed ignoring the external audit which is required for the products to be marketed internationally.

The concept of straight-forward warranty is being introduced to attract the customers in almost all electrical and electronic products. This can be further supplemented by the concept of contract maintenance or after sales service.

LIFE TESTING REQUIREMENTS

During production, devices are strictly controlled and screened to eliminate those with a potential for failure. In addition, inspections of product characteristics and periodic reliability monitoring ensure that the product quality level remains high. After product selection tests are carried out in conformance with the test standards.

Information collected at all stages of product development, from design to production, as well as from reliability testing as described above, is used to maintain and improve devices and to provide our customers with the safest and most reliable products possible.

Before testing there are number of points that must be considered when determining tests, stress conditions and sample size.

In order to satisfy reliability test objectives as discussed above, the following must be considered:

1. Where the product is to be used?
2. What kind of environment and operating conditions of the device are used?
3. What are the expected failure modes and mechanisms of the device, and what type of accelerated stress testing is appropriate?
4. What reliability standards are required for the product in the market?
5. What is the expected life span of the product?
6. How does the product rate in terms of motivation and importance?

Following are the accelerated stresses which can be applied on the products depending upon its type, for example on electronic items:

- Temperature
- Temperature and humidity
- Applied voltage
- Temperature cycling
- Current density
- Vibration
- Shock
- Constant acceleration
- Impact etc.

An important consideration in reliability testing is that test results must contribute to the evaluation and improvement of product reliability. It is therefore important to apply reliability testing consistently and repeatedly, to perform detailed failure analysis, and to carry all results back to the device design and processing stages.

SAMPLE SIZE SELECTION

Generally all the items produced should pass the quality and reliability tests before they are used in actual working conditions. When the products are manufactured in lots or in batches of a given number it is difficult and impossible to test all the items particularly when the lot size is large. A practical solution to this problem can be to take random samples from the lot and test each of them. With this procedure a risk factor either for customer or for producer is involved. When unsatisfactory lot is passed it will be called at customer's risk. Similarly when a good lot is rejected it is the producer's risk. The various sampling schemes used include the following:

1. ***Single Sampling Scheme:*** Here sample of n items from a lot size of N are tested.

Let c be the number of items to be defective. Then accept the lot if number of defective is less then c otherwise reject the lot.

The lot tolerance percent defective (LTPD) sampling is based on the customer's risk and acceptable level (AQL) is based on the producer's risk.

2. ***Double Sampling Scheme:*** Single sampling scheme can result in sudden death outcome where whole lot may be rejected on the evidence of a single sample. In a double sampling scheme, a second is taken in order to reinforce the evidence of first sample.

Here we test a sample of n1 items from a lot size of *N*.

If the number of defects less than c1 - accept the lot

If number of defects greater than c2 - reject the lot

If c1 less or equal number of defects less or equal c2 - take another sample

Then treat the operation as single sampling with sample size n1 + n2 and acceptance value c3.

The total number of item tested will be less than in a statistically equivalent single scheme.

3. ***Multiple and Sequential Sampling Schemes:*** The principle of double sampling can be extended to multiple sampling; here a large number of small samples, all of the same size n, are taken until a decision either to accept or reject the entire lot is reached. The advantage of multiple sampling is that the total number of items tested will be less than in statistically equivalent double scheme; the disadvantage is that it is more complex.

RELIABILITY TEST STANDARDS

Various industries and associations establish their own reliability standards depending upon the working environment of the equipment/system. Some of the standards for electronic components are listed below:

1. *Electronic Industries Association of Japan (EIAJ):* It has listed a few standards for reliability testing of semiconductors which specify Environment and Durability testing methods under various conditions.
2. *U.S. Military Specifications and Standards (MIL):* These standards are used for reliability testing of Electronic and Electrical Components and parts, such as semiconductor devices and microelectronics.
3. *International Electro Technical Commission (IEC):* These standards specify Environmental testing part-1 and 2 with mechanical and climatic test methods for semiconductor devices.
4. *Japanese Industrial Standards (JIS):* These standards specify environmental test methods for electricity and electronics.
5. *Joint Electron Devices Engineering Council (JEDEC):* It incorporates various test methods.

6. *CENELEC Electronic Component Committee (CECC):* These standards include general specification monolithic integrated circuits and general specification digital monolithic integrated circuits for testing electrical and electronic components.

TEST METHODS AND PLANNING

The method of reliability testing can be classified in two categories:

(*i*) *Test elements group (TEG)* — in which a special set of devices are created for the purpose of analyzing each failure mechanism, and product evaluation, whereby overall product evaluation are performed. TEG evaluation targets basic failure mechanism. A set of devices is manufactured especially for this purpose. This method allows detailed evaluation and analysis of failure mechanism. It is very effective for quantifying limits and allows use of accelerated conditions.

(*ii*) *Design approval testing (DAT)* — which eliminates failure mechanism that affect reliability in products.

Product reliability testing should approximate to the actual field environment and must always be repeatable. For this reason tests should be performed according to approved standards.

Similarly mechanical tests such as vibration, shock, constant acceleration etc. can be carried out at normal working conditions.

Accelerated Lifetime Tests

With ever increasing requirements of components reliability, the need to evaluate product lifetimes and failure rates is now greater than ever. Reliability tests simulate actual conditions of the products. Testing under normal/actual working conditions would take exceedingly long time. Therefore, stresses beyond absolute maximum rating are applied to the devices to

Table 10.1: Product Reliability Test Method Example—Semiconductor Device

Type	*Test*	*Description and conditions of test*
STESTEM ITEFIL	Steady state operation	Apply electrical stress (voltage, current) and high temperature for Extended period of time
	High temperature, high humidity baising	Apply electrical stress (voltage, current) high temperature and moisture for extended time period
	High temperature storage	Apply high temperature for extended period of time
	Low-temperature storage	Apply low temperature for extended time period
	High-temperature high - Humidity Storage	Apply high-temperature high-humidity conditions for extended period of time

accelerate the failures. In this way, device lifetime and failure rates can be determined, failure mechanism can be analysed. This type tests are called as accelerated life tests. Such tests are also used to force device degradation in order to determine a device's wear out period.

In general, if the degradation is fairly simple, life times and failure rates can be estimated relatively accurately using accelerated testing. Complicated failure mechanisms are difficult to simulate. It is important to select test conditions such that failure mechanisms are relatively constant, few and simple.

Constant Stress and Step Stress

The accelerated tests are of two types, constant stress and step stress. In a constant stress test, the time dependent failure distribution of a sample subjected to constant stress at several stress levels is observed. In a step test, stress is applied to a test sample gradually in stepped increments, and the step which failure occurs is observed.

Temperature

Accelerated life testing is closely related to the physics. The physical and chemical reactions inherent in device degradation are often treated as examples of chemical kinetics. Chemical kinetics is a basic chemical reaction model that describes the temperature dependence of failures. It is seen that the temperature dependence increases as the activation energy increases.

TEMPERATURE AND HUMIDITY

Moisture Resistance Tests

The reliability of some of components depends largely on the moisture contents during their working for example, semiconductors. Various types of moisture resistance evaluation tests are developed in order to evaluate these devices quickly.

While testing the devices sometimes they are subjected to humid atmosphere or the moisture is allowed to penetrate the device.

Voltage

Aluminum corrosion due to humidity and characteristic degradation due to mobile ions are accelerated by voltage. In these cases, voltage is applied continuously to the oxide film and breakdown occurs over time, even if the electric field in oxide film is below the breakdown limit.

Temperature Cycling

Temperature cycle thermal shock tests verify reliability against repeated thermal stress from external or internal heat sources. In the test the device is subjected to low and high temperatures repeatedly. Failure modes such as bonding opens, passivation cracks, package cracks, chip cracks and characteristic degradation (due to piezo effects) are checked. In the temperature cycle test the device is usually subjected to it's high and low temperature limits repeatedly. In some case these limits are exceeded to achieve even greater acceleration. However, it should be noted that, because of the different materials used, different areas of device will have different physical characteristics and will produce varying test results. Therefore, the behaviour of the device in the field may not match the test results.

FAILURE RATE ESTIMATION

It is important from the point of view reliability and safety to estimate failure rates for the components used in different equipment.

Failure rates are estimated by calculating an acceleration coefficient based on an acceleration life test or by gathering failure rate statistics from products used in field.

Estimation Using Accelerated Life Tests

With rapid technological change in industries, accelerated life testing has become increasingly important as a practical and economical means of evaluating reliability. This is essential to bring the new products to the market quickly. Here it is described how failure rates can be estimated by the use of accelerated life test in the field.

When estimating failure rates from test data in the field, the actual number of failures is often very small or, in many cases, zero. In such cases, the failure rate must be estimated on a certain confidence level. When the failure pattern follows exponential distribution the following relationship can be used.

Failure rate $\lambda = r/T$ where,

R= number of failures; T = total components hours

RELIABILITY TESTS

The reliability tests are required to show that the finished products meet the reliability specifications defined during design stage. The target failure rate within the limits under all specified environmental conditions must be verified. A full reliability test may comprise of basic three activities and include:

(*i*) **Failure rate:** Under specified conditions of temperature range, stress and environmental conditions the failures of the product must be measured on individual items or on random samples. The test may involve, testing at different level of temperature, humidity, mechanical vibration, mechanical shock, operating voltage and salinity.

(*ii*) **Burn-in:** The above cited conditions of testing are made more severe than the defined specifications. This is done to accelerate the failure of weak or defective components so that they fail during test. The aim here is to eliminate or minimize the early failure region as shown in bath tub curve, so that the useful life of the product is extended. This is normally done for electrical and electronic components at high level of temperature humidity and voltage.

(*iii*) **Screening:** This includes visual inspection constant acceleration and measurement of electrical parameters of electronic components and assemblies.

IMPLEMENTATION OF RELIABILITY TESTING

Design Approval Test Procedure

Reliability tests on the devices are performed in research and development phase and in full production phase. During research and development, reliability tests are used to evaluate the quality of design, the material and processes. During full production they are used as design approval tests (DATs) and for periodic reliability monitoring.

During prototyping of the product the components are divided into families according to their performance. Reliability testing is performed on a representative product from each family. Tests for early failure rate, (EFR), long-term life or random failure rate (IFR), time to wear out and resistance to environment are performed.

Reliability Monitoring During Full Production

Products passing DAT are given an initial quality assurance inspection before shipping. In addition periodic monitoring is performed to verify the reliability levels of shipped products. In the initial quality assurance the basic features and appearance are checked. The objectives are to verify the product quality of each product lot and to assure the quality of the shipped products.

Periodic reliability monitoring involves life and environment tests performed on groups of products classified by process and package family. Reliability is continuously monitored, failure rates are analyzed and results are fed back to the manufacturing process. In addition, data are accumulated allowing reliability to be further improved.

For any type of reliability testing, test set-up are designed to create actual working conditions. The loading range of the products plays an important role to indicate the failure mode.

Data Reporting System

Reliability testing is performed on product families during product development and periodically thereafter to monitor reliability. The results are given to the customer at three-month intervals in the form of reliability reports—Referred as Quarterly reports—which contain the results of the tests and the results of initial quality assurance inspection for each family.

Data Reduction and Analysis

For mass production system the quantity of the data can be very large, which can not be used directly for reliability evaluation, therefore selection of a sample size is needed. This can be done through data reduction and analysis.

QUESTIONS

1. Explain, why reliability testing required.
2. How sample size for reliability testing is selected?
3. Describe various methods used for reliability testing. Why their planning is required?
4. Discuss various reliability test standards used in practice.
5. How failure rate can be estimated? Explain.
6. Enlist the steps required for the implementation of reliability testing.
7. How does data reduction and analysis helps in reliability testing?

11

Maintenance Information Systems

INTRODUCTION

The use of highly sophisticated equipment in the modern industry calls for electronic and computerized information handling to allow the effective maintenance management of this type of equipment. In case where advanced manufacturing technology is being practiced the use of information technology will provide the following:

- Control over production machinery.
- Evaluation of system performance, through management information system.
- Integration of equipment to use as a system rather than a single unit.

Information technology can be described as the mechanism by which instructions concerning production, maintenance etc. may be accurately encoded, transmitted, received, stored, processed and used. This can help in generating expert system for automated production as well as maintenance. The concept of information technology should be applied at each level of control within the system, as well as, to the external production, maintenance and management functions, which support the equipment/facilities. With, this a network is formed for the control of day-to-day activities in an organization.

The failures of automated production system can cause substantial loss of production, if the standby facilities are not made available. Thus, it is pertinent to give more emphasis on maintenance functions to ensure operational availability and continued operation of such equipment. Some managers do not appreciate the investment on maintenance skills, equipment and type of maintenance policies to be followed, which directly contribute via high plant availability and better product quality.

ROLE OF MAINTENANCE INFORMATION SYSTEM

The importance of maintenance information system (MIS) increases many folds with the utilization of modern sophisticated production plants. The improvement in technology and use of automated equipment, the role of maintenance engineers calls for proper attention towards maintenance functions to improve the system/equipment availability and reliability. For improving the availability/reliability of equipment/system, it is desired that the information in connection with the failure data be made available, which will help to find the failure rate and failure density functions. This information can further be used to assess the system reliability. The main role of the maintenance group has to be to maximize the mean time between failures (MTBF) and minimize mean time to repair (MTFR). For effective planning and implementation of any maintenance policy, it is therefore, desired that the authentic and complete information must be made available.

OBJECTIVE OF MIS

The objectives of developing any information system are two fold (*i*) to provide the type of information that will help better coordination, and (*ii*) to provide management with access to information relating to the activities undertaken by the group in any complex organization.

Through reports, the management can be informed about the activities and limitations, if any, in any maintenance department of an organization. These reports will provide the following information to the management:

1. Information required taking day-to-day decisions.
2. Preparation of database for formulating policies related to maintenance/production functions.
3. Analyse and measure the performance of different inputs i.e. men, material and machinery, etc.
4. To know-how effectively the various divisions are working in the organization.
5. Point out the deviations from the pre-set standards and their rectification if required.
6. Identification of various cost components, which are not included in direct, cost measurements.
7. Point out wastage of material, time and process loss, if any.
8. Initiate and implement cost effective measures to justify extra cost spent on information system.
9. Assist in maintaining the quality of work to enhance the operating life of the equipment/ system through better maintenance practices.
10. Improve the existing situation by reporting favourable or unfavorable trends, in the organization.
11. Motivation of working personnel for effective integration of their potentials to meet out the organizational goals.

DESIGN OF MIS

Before designing any maintenance information system, it is necessary to plan it before hand, but there is hardly any way, which can be best for system plan. The important principles are given below:

1. The planner must acquire sufficient knowledge of the work undertaken by him/his team.
2. The users also must be made a party in the planning process wherever possible.
3. The experts in the field should assess alternative planning approaches through group examination and discussions. After proper evaluation only the final plan be accepted and adopted.

The actual design of maintenance information system is discussed in the following paragraphs.

Planning Process

During planning process, the following issues must be considered which are important from the user's point of view:

(*a*) *Mechanised vs Non-mechanised Information System:* Use of computers is only made when adequate gains can be foreseen. In general, there are advantages of the mechanized system which include reduction in labour cost, improved processing speed, ready access to information and reduction in expenditure to store, analyse and organize the massive data. The only disadvantage is the initial cost which can be compensated with use of information system over a period of time.

(*b*) *Centralization vs Decentralization:* The centralized information system has better control over all the functions and greater ability to share information as compared to decentralization system. Here again with the use of computers it is possible to use the centralized information system more effectively.

(*c*) *Batch mode vs Real Time Mode of Information Processing:* The batch-processing mode visualizes easy availability of data from the source at the point where such data are being processed. Whereas, on-line real time processing assumes that complete information is available as input device (to date) at the point where users are operating. However, the choice of a particular system would depend upon the planner that how fast the information can be communicated to the users in the desired format.

(*d*) *Dependent vs Modular:* This largely depends upon the area of application. It is dependent when showing of information is restricted beyond certain area, however, under modular system; the application areas are divided into several modules. Some of which may be in common with other areas. The planner must consider the cost benefit analysis of the system suggested by him and the apportioning of cost should be done carefully based on the man-hour spent and computer hour consumed.

Design

The design of information system is a creative and time-consuming activity. Some parts of it can be automated but most of the creative aspects require human thought. The system design, therefore, is centered around establishing the information base in any organization and can be divided into three stages: (*i*) assessing feasibility of information system (*ii*) designing information base, and (*iii*) organizing access to maintenance information system.

It starts from the complete evaluation of existing system. This is undertaken in order to define the application area along with its extent and complexity and also to provide scope for the use of the computers. The input and output formats must be defined. In order to establish that the proposed system is feasible the criteria applied may include (*a*) ability to meet users and (*b*) use of resources to generate profits (*c*) impact of information system on the user organization and (*d*) workability of the proposed system.

The design of the information system should be such that all input and output either, must be directly from input, from a file or from computation based on the input and file data. With the requirements of outputs, the contents of the files and input can be determined and users have to design the input methods only.

The files contents and structure must be designed in consultation with the users who will be in a better position to indicate the size of data items and to specify updating needs. The type of system needed will largely depend upon the volume of input/output activities and updating needs.

In other part in feasibility study includes the cost analysis of computerized and manual maintenance information system. The impact of new information system on the organization as a whole can be tested to determine, the benefits achieved, simplification of work procedure and sharing of duties and responsibilities. The next step to be examined will be the workability of the system in the existing environment. It is well known that initially any new system faces criticism from the users, and it has to justify its existence in the present environment, which is likely to be changed.

IMPLEMENTATION OF MIS

The task of management of maintenance in the modern environment is a difficult task but important because, through it both effectiveness and profitability are achieved by the organization. The problems associated with the management of maintenance system include, system availability, reliability and maintainability of hi-tech equipment/systems being used in the modern industries. It is important to understand that high plant/system availability is absolutely dependent on the inherent reliability and maintainability incorporated in an equipment/system design. The initial system's availability is governed by the above two parameters based upon the equipment's functional design. But subsequent plant/equipment availability is directly dependent upon how it is being maintained. With passage of time the equipment/system will tend to degrade with normal usage.

It is observed in practice that high availability is essential in respect of flexible manufacturing system (FMS). The problems associated with such type of system forces the users to adopt on-line condition monitoring for the FMS and also the possibility of utilizing data so generated within its MIS for management control. This example shows the basic need of maintenance information system. In this case data were collected by several computerized acquisition and analysis units, fitted with FMS and linked with mainframe computer, whereas, other information related to products schedule and shift patterns etc. are also made available as and when required.

A technique, which can help at the specification level, to outline the maintenance organization and identify information required at particular points, is the system modeling and analysis tool (SADT). The fundamental benefit of SADT is that it can be used to model any system irrespective of its complexity, considering overall objectives of the system.

Since the database is the key to a successful interactive maintenance system, it must support the achievement of the objectives of the maintenance of system, must interface with existing/future systems and must be supportable and expandable in a distributed processing environment. The effective operation of maintenance system is based on the type and amount of data stored in its memory and the system efficiency will go down when there is no correct exploitation of this data. SADT can be successfully used to create the structure of an interactive database, to support the maintenance system, via the modeling of information input and flow. To access the information correctly specialized language programs can be used.

EXPANSION OF MIS

In the competitive industrial climate, use of highly automated and flexible production equipment is inevitable. To achieve high level of competence, the techniques of information technology should be correctly used to manufacturing operations. Under above conditions, the role of maintenance department of the organization has a significant contribution for maintaining the equipment/system under their control to utilize the fully automated system effectively. The areas of repair technique, logistic support, organizational structure, work practices, staff skills and training are to be given importance. This is only possible through proper implementation of Information Technology to all related areas. Thus, failure prevention becomes the prime requirement of the future maintenance department, besides the routine nature of maintenance, i.e. breakdown maintenance and time based maintenance.

If the requirement of an automated system for production is high i.e. round the clock, the use of computer based maintenance system (CBMS) would be desired to maintain the equipment availability to 100%. This will be only possible through information technology, together with the concepts of artificial intelligence (AI), to integrate the condition monitoring (CM) and maintenance information system (MIS). With the use of above methods, it will be possible to provide optional, expert and predictive management control.

Analysis of information is the fundamental process, to assess where on-line maintenance system can be employed. Therefore, CBMS via integration of predictive maintenance and maintenance information system can be employed for automated production system, which basically relies totally upon the actual condition of the equipment/system.

The proper utilization of fully automated production systems can only be made financially viable with their better availability. This in turn, can be helped by integrated maintenance information system. To maintain such equipment/systems highly trained staff will be required to operate computer integrated production facilities, if a reasonable degree of equipment availability is required.

EVALUATION OF MIS

In practice, majority of maintenance information software's are available but they require certain changes over time, to meet new circumstances and/or extra requirements of any new system. In this way, the information system, which meets the requirements of the organization, will be considered dynamic. Accordingly, while developing MIS, due consideration should be given to the many initiatives, which are needed for the modern maintenance practices. For example, basic knowledge of terro-technology can help in building up a new maintenance information system, which can minimize life cycle cost and other related functions. In recent years, the advent of better data storage, speed of retrieval and use of maintenance information system or expert systems, has allowed many new ideas or techniques in maintenance to be implemented, which include condition monitoring, logistics engineering, total productive maintenance, asset management and reliability centered maintenance.

The common link between the above techniques or methodologies in maintenance is the acquisition, manipulation, analysis and interpretation of data. With the use of maintenance information system, it is possible to use the condition monitoring techniques to assess the intelligent computerized equipment where the services of the operator can be taken to carry out first time maintenance with due training for effective utilization of the manpower. Sensors are being used to monitor the condition of the equipment but they should be accurate, intelligent, simple and highly reliable without increasing the system complexity.

The maintenance information software's associated with such sensors will be very powerful for the purpose of diagnosis of the system faults/failures and will also provide fault reporting repair control, labour reporting and control, equipment spare part inventory control and manufacturing date recording and analysis.

QUESTIONS

1. Discuss the role of maintenance information system.
2. What are the basic objectives of maintenance information system? Explain it.
3. Elaborate how maintenance information system should be designed for effective use.
4. Highlight the problems faced during implementation of maintenance information system.
5. Explain how existing maintenance information system can be expanded for better utilization in an organization.
6. Enlist the steps required for evaluation of maintenance information system.

12

Maintenance Concepts

INTRODUCTION

It defines the repair/maintenance methodologies, its support levels, personnel factors, time constraints etc. and serves two way purposes:

(*a*) Establishment of maintainability requirements which should be part of design of products. For example, if external support for testing is not provided the same should be built within the system.

(*b*) Establishment of requirements for total maintenance support including manpower with skill and other tools equipment required.

Prior to the start of actual design the above two concepts must be developed for effective and economical fulfillment of the maintainability objectives.

A machine, equipment/system, once designed and fabricated is bound to fail during course of its operation, which can be delayed or avoided by regular maintenance. The basic objective of any organization will be to keep their equipment/system operational to achieve the goals of the organization. Measures taken by the industry to keep machines and operating systems in trouble-free condition are collectively termed maintenance engineering. Successful maintenance depends upon a sound management structure together with robust diagnostic techniques with the adequate planning and control of activities. A high level of automation, deploying complex and sophisticated machines, characterizes modern industries, therefore, the role of maintenance personnel has increased many fold and new maintenance practices are being used.

The idea of maintenance is very old and was introduced along with the inception of the machine. In the early days, machines were used, as long it remained operational. When it stopped working, it was either repaired or discarded. In today's world, the high cost automated machines/equipment need to be properly maintained/serviced during their entire life cycle for maximizing their availability.

Maintenance is a function to keep the machines/system in working condition by replacing/repairing some of its components/parts. Sometimes even periodic inspection is sufficient to keep them running. The maintenance concept is an outline plan of how the maintenance function will be performed. With the information available from users, detailed procedures are drawn to concretize the maintenance concept. The executable procedures developed, these are collectively called the maintenance plan. The development of such a maintenance plan is one of the most important requirements of the maintenance program that requires interaction between the user and the manufacturer, which can also suggest required modifications in design and material.

The maintenance concept is closely related with the operational needs of the equipment/system. First of all, the operational concept is determined and then the maintenance concept is developed to support the operational concept. Finally, a maintenance plan is prepared in accordance with concept thus developed.

The term 'terro-technology' is directly associated with the maintenance and is concerned with the specification and design for reliability and maintainability of equipment, machinery, etc. with their installation, commissioning, operation, maintenance, modification and replacement, with feedback of information on design, performance and costs.

WHAT IS MAINTENANCE?

Maintenance has been defined as (BS 3811.1984):

The combination of all technical and associated administrative actions intended to retain an item in or restore it to a state in which it can perform its required function. This required function might be defined as stated condition.

Maintenance is the routine and recurring process of keeping a particular machine or system in its normal operating condition so that it can deliver its expected performance or service without causing any loss of time on account of accidental damage or breakdown. In other words, maintenance means, the work that is required to be done to keep equipment in a running condition such that it can be utilized to its full designed capacity and efficiency for maximum amount of time.

With the technological development all around the responsibilities of the maintenance engineers have increased considerably due to over growing complexity and size of industrial organizations. It has become very important to make effective use of available facilities that have been set up at high investment costs. To achieve maximum profit, it is essential to run the equipment/system effectively and it is only possible when the equipment/facilities are looked after properly. Maintenance function also involves safety aspects of certain equipment where failure of a component may cause a major accident.

Maintenance is also related with the profitability through equipment output and its running cost. Maintenance work improves the equipment performance level and its availability but adds to its running costs. The objective of the maintenance work should be to strike a balance between the availability and the overall running costs. The responsibility of the maintenance function should, therefore, be to ensure that production equipment/facilities are available for use for maximum time at minimum cost over a stipulated time period such that the minimum standard of performance and safety of personnel and machines are not sacrificed. With this in mind now days, separate maintenance departments are formed in all industrial organizations to look after all maintenance requirements of equipment and facilities.

PRODUCTIVITY AND MAINTENANCE

Any equipment/facility designed and fabricated is with certain purpose of providing desired output in terms of goods or comfort to the mankind. It is seen from experience that the output from any facility is directly dependent upon its up keep. Breakdowns/failures are inherent part of any engineered product. These breakdowns can be minimized with the help of proper maintenance functions. For achieving the set goals of production, the equipment/machine availability must be high. The plant availability can be described as:

$$A = \frac{UT}{UT + DT} \qquad ...(12.1)$$

Where, A is the availability of the machine over a specified time period.

UT is machine's up time, and

DT is the machine down time during the specified time.

The up time of the machine/service is thus, the time for which it is actually available to complete the desired function and down time of the machine is the time period for which it is not operational.

The above definition of availability considers only two states of the equipment or machine i.e. working and failed. However, in actual practice, there will be spectrum of intermediate states as well. For practical purposes, idle (or ready) time of a machine or service as well as the total time lost in administrative or operational delays cannot be overlooked. Thus, it is more appropriate to express the availability of equipment in terms of its working effectiveness called operational availability (*OA*) expressed as:

$$\text{OA} = \frac{OT + IT}{OT + IT + AD + RT} \qquad ...(12.2)$$

Where, OT is the operating time

IT is idle time

AD is administrative or operational delays

RT is the repair time.

MAINTENANCE CHALLENGES

The maintenance function of a modern industry faces a number of challenges attributable to:

- Rapid technological growth.
- Advent of new advanced diagnostic tools, rapid repair system, etc.
- Advanced store management techniques to incorporate modular technologies.
- Availability of maintenance management information system to expedite maintenance related functions.
- Requirement of keeping both outdated and modern machines in service for effective production.
- System design complexity to incorporate many functions in one equipment/component.

The effective management of maintenance aspects under such challenging circumstances is often difficult. Besides diagnostic of fault in the equipment, the activities of the maintenance department include:

- Upgradation of the existing plants and equipment and training of maintenance staff to attain the required technical skills.
- Effective maintenance of the old equipment for higher availability with minimum expenditure.

- Cost optimization of all maintenance functions to minimize overall costs.
- Improvement of maintenance activities in the areas of tribology and terrotechnology.
- Reconditioning of used/unserviceable spare parts, in case of non-availability of genuine spare parts.
- Development of indigenous sources for spare parts for import substitution.
- Setting up of an effective maintenance information system (MIS).
- Setting up of in-house R&D activities for effecting improvements in maintenance practices.

The above requirements can only be possible with setting up an organized maintenance department with trained and qualified personnel.

MAINTENANCE LEVELS

The maintenance functions can be performed at various levels which include the followings:

(*i*) **Organizational level:** Here the maintenance/repair work is carried out within the capabilities available in the organization and all related facilities must be made available. Normally at this level the maintenance is limited to periodic checks of equipment performance and minor repair works.

(*ii*) **Field level:** In this case maintenance functions are performed by a mobile or semi mobile_team and are limited to the repair of end items and the other items which cannot be shifted to the workshop.

(*iii*) **Field installations:** These are generally established to support lower levels of maintenance work within the specified geographical areas. The tasks which cannot be done by mobile units are undertaken by such establishments. High personnel skills, additional test facilities are provided at this level.

(*iv*) **Maintenance workshops:** This constitutes highest level of maintenance and provides the support for the works which cannot be undertaken at other levels. Here bulky and complex equipment for maintenance functions are deployed and the concept of assembly line techniques is practiced. It provides facilities for complete overhauling and re-building of equipment/ system.

The facilities of testing, calibrations etc. are provided at these workshops. This type of system utilizes its facilities more effectively where similar type of equipment system are used at different places.

MAINTENANCE PLANNING AND SAFETY

The complexity of equipment and overall technological developments is compelling the manufacturers to be safety conscious. Legislations and government directives also force them to comply with the requirements of safety aspects in their products. The challenging concept of safety is to be incorporated at the design phase of the product. Human safety has therefore gained maximum attention of the manufacturers. The objectives of safety program include the following:

(*i*) The safety should be incorporated at design level which must meet the mission requirements.

(*ii*) Failures associated with the product (components/units) should be identified, eliminated or controlled to acceptable level.

(*iii*) Emphasis should be exercised to minimize, eliminate failures to protect equipment/system.

(*iv*) To obtain desired safety, use of new material/processes must be minimized.

The planning program for safety identifies specific tasks at various phases of the product which are:

(*i*) **Formulation phase:** During this phase studies should be conducted to carry failure analysis and define product safety performance and select safety effectiveness measures.

(*ii*) **Definition phase:** Here the safety aspects must be planned and implemented. Identify safety decisions to be made before the development phase.

(*iii*) **Development phase:** The program approved during definition phase must be implemented with its evaluation through failure analysis and safety studies.

(*iv*) **Final phase:** Maintain safety precautions throughout including audit of engineering changes.

Safety aspects are also important beside human beings and include industrial capital assets. It is true in case of organizations involved in production of toxic material and other dangerous substances operating under severe conditions. Here the failure will cause not only loss of human life but also the material. One example of it is the Bhopal (India) gas tragedy about 20 years back. In the above case the failure was caused due to negligence of the operator and the maintenance personnel. Different organizations have different degree of risk depending upon the nature of product, the raw materials and the processes involved. Therefore along with reliability of the products/process the safety aspects are also critical and important.

MAINTENANCE OBJECTIVES

Like other organizational objectives, the maintenance objectives must be well defined to maximize the equipment availability to meet out the organizational goals. The main objective of maintenance is the establishment of safe working conditions both for maintenance and operating personnel. Fig. 12.1 below shows the other important objectives of maintenance functions.

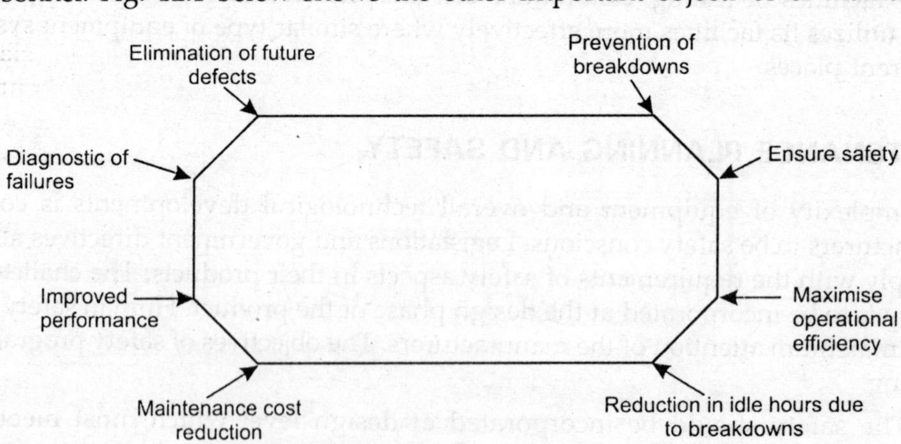

Fig. 12.1: Objectives of Maintenance

In the present industrial organizations, the importance of maintenance functions has drawn the attention of the users. Therefore, its objectives should be formulated within the framework of the overall organizational set up so that its goals are fulfilled. The maintenance plans can be drawn up to provide guideline within which the maintenance actions can be carried out. For this, the maintenance executives need to ensure that:

(*i*) The equipment or facilities are always in an optimum working condition to minimize power or fuel requirements.
(*ii*) The time schedule of delivery to the customers is not affected because of non-availability of machinery/service.
(*iii*) The performance of the machinery/facilities is dependable and reliable.
(*iv*) The downtime of equipment/system is kept to minimum in the event of breakdown.
(*v*) The maintenance cost is properly monitored to control the overhead costs.
(*vi*) The life of the equipment/system is prolonged while maintaining the acceptable level of accuracy of performance to avoid unnecessary replacement.

One of the important and vital objectives of the maintenance function is to keep the maintenance costs under control throughout the life cycle of the equipment. If the cost of the maintenance/operating cost increases to a great extent because of non-availability of spare parts leading to consequent delays in maintenance works, it is always advisable to bring this to the notice of the management and propose the procurement of new equipment that can be used in place of existing one. The important merits and demerits of new equipment must also be pointed out. This will help the organization to formulate its replacement policy for modernization of the plants.

System Approach

Once the objectives of the maintenance department are set, the system approach to maintenance involves the following:

(*i*) Identification of the needs of maintenance related activities, which may include repair, reconditioning or replacement of components.
(*ii*) Analysing the requirement of the above needs.
(*iii*) Determining the functional procedures for maintenance task selection, work planning and scheduling, work order processing, etc.
(*iv*) Outlining a reporting and controlling procedure of all maintenance related activities.
(*v*) Development of supporting services and infrastructure for efficient execution of maintenance functions.
(*vi*) Determining the cost account procedures for optimizing maintenance related expenditures.
(*vii*) Adopting a policy for training of maintenance staff and assurance of quality.

Thus the system approach is applied to ensure better maintenance performance for a particular environmental situation.

MAINTENANCE REQUIREMENTS

To meet out the challenges and objectives of the maintenance function effectively, the maintenance requirements must be examined carefully, which may include the following:

- Development of maintenance plan for planned as well as breakdown maintenance.
- Development of maintenance structure/facilities to carry out maintenance work effectively.
- Development of spare parts control mechanism to keep the down time of the equipment at minimum level.
- Planning and control of manpower for maintenance functions.
- Optimization of maintenance costs, for recommendation of a particular policy.
- Feedback and information system to develop strategic planning for future requirements.

The first and the foremost task of the maintenance management team should be to prepare a contingent maintenance plan of all the equipment/system under their control. Depending upon the criticality and need of equipment, a particular policy of the maintenance function must be adopted keeping in mind the cost parameters. The next step would be, to develop the infrastructure and facilities to carry out the maintenance functions as designed.

The role of spare parts management is very important and crucial to reduce the downtime of equipment/system. This will only be possible if the required spare parts are made available in time. Maintenance function is a most unpredictable activity, therefore, the assessment and planning of manpower is very difficult. Due care should be taken during manpower planning for maintenance function so that optimum utilization of the available manpower is made. Extra manpower will cost more and at the same time, their shortage will make the maintenance work to suffer.

Economic aspects of the maintenance functions are important and before adopting any maintenance policy a clear analysis must be carried out, to supplement the expenditures made for the maintenance function. It is a general feeling in the organization that the maintenance cost is the burden on the production; therefore, efforts must be made to minimize it without sacrifying the availability and reliability of the equipment/system. At this point of time, it is pertinent to note that a feedback information system must be developed to keep the track of all the activities being performed by the maintenance department. This will help even the manufacturers to modify their product to meet out the challenges of the growing technology.

MAINTENANCE ACHIEVEMENTS

Effectiveness of the maintenance depends upon systematic conductance of various interrelated activities. To achieve the goals of maintenance system, certain established principles must be followed. These principles help and guide the maintenance staff in efficient discharge of their duties. The main areas of work governed by these principles are discussed below:

Plant Management in Maintenance

The primary duties of the maintenance group include safe and effective operation of equipment/ system, to achieve the desired production targets on time and economically. Sometimes it is observed that scheduled maintenance is not carried out, since the production department does not release the equipment to meet out their production targets. Therefore, it is necessary that all the departments cooperate with each other to meet the organizational goals effectively.

Plant Objectives vs Maintenance Objectives

The objectives, of the plant must be formulated considering the maintenance objectives to avoid clash so that the maintenance work is carried out in time without disturbing the production targets. If this is not practiced, may result in a situation, where both the work i.e. maintenance as well as production is bound to suffer.

Documentation

The preparation of the work order for the maintenance work is an important task, since it indicates the nature of the work to be performed and the series of operations to be followed for a particular job. It is essential that at every workstation, proper records/enters must be made to monitor the health of the equipment as well as its life span. Now a day's use of computers can be helpful in this regard.

Decision-making

For successful completion of any work, the importance of a reliable information system cannot be overlooked. The information collected can be used for decision-making in respect of manpower and spare parts requirements; therefore, data entries for any type of breakdown must be recorded in time, so that the same can be used as and when required.

Adherence to Planned Maintenance System

It is noticed that the instructions of the manufacturers are not given attention while the equipment/machines are in use. This often gives rise to unscheduled breakdown of machinery. If the instructions of the manufacturer or even planned maintenance schedule developed at the organization level are followed strictly, will not only enhance the operating life of the equipment but will also ensure proper use of maintenance manpower.

Planning Maintenance Functions

Minor maintenance/repair jobs can be done without any proper planning but major jobs must be planned in advance for effective utilization of manpower/facilities and to obviate incurring of heavy shutdown costs because of equipment breakdowns.

After completion of each major maintenance job, the cost incurred and the performance obtained must be compared with standard figures for justification.

Manpower for Maintenance

Manpower planning for maintenance function is a tedious job since the requirements are varying because of unforeseen breakdowns. Therefore, estimation of manpower must be based on the facts and figures of maintenance activities. For the purpose of scheduled maintenance work, it is easy to make the assessment as the work contents are known in advance whereas, for random breakdowns it is really difficult. But based on the past experience calculation must be made for completing all types of maintenance work. The control of the work force is also essential to control the maintenance cost. This can better be done, when relevant information is made available to the users.

Quality and Availability of Spare Parts

Some quantity of spare parts is supplied by the manufacturer at the time of procurement of the equipment but may not be sufficient in due course of time to meet out all maintenance requirements. Therefore, it is desired to maintain an inventory of complete range of quality spare parts to sustain the functionalities of the equipment/system without endangering safety.

With changing pace of technology, the nature of maintenance function also needs to undergo changes. Whenever, a new equipment/machine is purchased, the maintenance personnel along with the operating personnel must be trained. The supervisory staff also requires appropriate training to ensure quality of maintenance work. On job training of new recruits under skilled workers and supervisors is being practiced in many organizations. Evaluation of training is also essential to meet the growing challenges of maintaining sophisticated and delicate equipment/systems. These days, formal training is also being provided by some educational institutions in form of short-term courses to meet out industrial requirements.

R&D MAINTENANCE

In the Indian industrial organization, the concept of maintenance to improve and to enhance the life of the equipment is growing fast. Lot of studies has been carried out in the developed countries to improve the reliability of the equipment/systems, and the concept of maintainability is also gaining its significance to make maintenance job easy. But the following areas need attention of all concerned:

1. *The influence of maintenance on design and selection of plants:* Here, how designers, users and manufacturers can exchange their information in larger interest of each other, so that the reliability of equipment/systems is within the specified limits and the maintenance/operating costs are also kept minimum.
2. *Use of available information:* How the information published on reliability and maintainability can be effectively used to improve the performance of the systems operating under terrane conditions.
3. *Training Requirements:* To make full use of the published information, it should be examined with care that what reliability and maintainability training and education is necessary amongst the practicing engineers and technicians to derive its full meaningful advantage.
4. *Standardization of Product:* Efforts should be made for the standardization of equipment particularly when it is imported, so that minimum problems are encountered during their maintenance. Inter-changeability of parts must also be given due consideration to make its use feasible.
5. *Spare-Parts Management:* It has been observed that the determination of correct quantity of spare parts is a difficult task, besides its procurements. R & D groups are formed to examine the problems related to the availability of genuine spare parts at a reasonable price. Here, also practicing technocrats can be educated through short-term courses to tackle the spare parts problem being faced by the industries.
6. *Development of Maintenance Facilities:* With the technological developments, the size and volume of the equipment/systems are growing at a faster pace. This calls for their timely maintenance to keep their downtime as minimum as possible. For the effective working of maintenance groups, special facilities required must be developed to face the challenges in the area of maintenance.
7. *The Concept of Condition Based Maintenance:* Attempt should be made to incorporate the modern maintenance practices on costly and sophisticated machinery/systems to improve their operational availability. The concept of condition-based maintenance is quite new

and helps in maintaining the machines effectively. Though its initial cost is high, but can be compensated with the better availability of the equipment and safety to the users.

8. *Concept of Contract Maintenance:* As indicated earlier, the maintenance function is complex in nature and needs services of the highly qualified and trained personnel. The concept of contract maintenance can help in this regard to a greater extent, since the services of the experts can be easily hired. Here, maintenance groups work for a particular product and definitely have expertise in their work.
9. *Development of Indigenous Spare Parts:* In case of imported equipment/systems, the problem of spare parts is really very acute and efforts should be made for its substitution at organizational level by developing technical know-how and substitute of material. The problem becomes more serious when the manufacturer stops the production of a particular type of equipment and the equipment is still in use at certain places. Obsolescence is another factor, which calls for development of technology to meet out the challenges.

ROLE OF OVERHAULING IN MAINTENANCE

The useful life of equipment is restricted for reasons such as deterioration with age or obsolescence. The equipment used in industry can be categorized mainly as (*i*) equipment with diminishing efficiency and (*ii*) equipment with constant efficiency. The first category of equipment deteriorate with time resulting in increase of operating cost whereas, second category of items operate at a constant efficiency for certain period of time and fail suddenly.

When the concept of overhaul is applied to the equipment with diminishing efficiency, the model used maximizes the gain between the operating costs before and after the overhaul is carried out to improve equipment performance and thus reduction in operating costs. It is assumed the effect of overhaul is a downward shift, that is, improvement in operating cost parallel to initial path. It is shown through study that a substantial amount of gain can be obtained in the operating cost of the equipment. The operating cost included electricity charges, maintenance charges, overhaul cost, charges due to failure etc. for the study conducted on Air conditioners. It was also pointed out that at what interval of time the overhauls should be carried out to achieve the desired gain in operating cost.

EXPERT SYSTEM IN PREVENTIVE MAINTENANCE

The initial cost of numerically controlled machine is quite high and therefore, enough precautions are made to ensure its smooth operation for maximum possible time. Moreover, unexpected equipment breakdown incur significant financial losses in manufacturing systems which are operated on highly optimized production schedules. In order to avoid unscheduled breakdowns concept of preventive maintenance with expert system can be used.

An expert system is a program which has a wide base of knowledge in a restricted domain, and uses complex inferential reasoning to perform tasks, which human experts could do. The important parts of the system are:

Knowledge Base: It encompasses knowledge about the domain problem. Knowledge representing methods are, production rules, frames, semantic nets and predicate logic.

Inference Engine: The inference engine is the problem solving control strategy or rule interpreter, which applies the knowledge in the knowledge base of the problem. The strategies are forward and backward chaining.

Two fundamental modules are supporting the system construction. They are (*i*) explanation system and (*ii*) knowledge-base management system.

Three interfaces complete the architecture of an expert system. They are user interface, knowledge engineer interface and external interface.

Development of Expert System: The system can be developed in high level languages like prolog, LISP, etc. But here one has to put a lot of efforts in developing the program to make it user friendly. It is very easy to develop an expert system by using expert system shell available in the market.

Validation of the System: The system is evaluated by the empirical approach. The approach is to test a representative set of selected test problems. The explanation system facilitates easy verification for the correctness of the expert system by an expert based on a comparison with his own decisions. This expert system provides guidelines for the maintenance staff. The periodic preventive maintenance that should be provided for the NC machines, has been kept in the knowledge-base of an expert system shell. The expert system gives some recommendations as per the user's input and also comments on the maintenance.

QUESTIONS

1. Discuss the basic maintenance concept applied to any industrial organization.
2. Define maintenance and how it is related with productivity.
3. Explain the challenges faced by the maintenance manager.
4. What are the main objectives of the maintenance? Discuss.
5. How system approach can be applied to maintenance functions? Explain.
6. Explain what are the maintenance requirements.
7. Enlist the steps required to achieve the required maintenance standards.
8. Discuss the role of R&D in maintenance organization.
9. Explain the various levels of maintenance functions.
10. What are the safety requirement in product design? Discuss.

13

Maintenance Practices

INTRODUCTION

As mentioned in earlier chapter that the mean time between failures (MTBF) be enhanced and mean time to repair (MTTR) must be minimized. This can be made possible through better maintenance practices. The introduction of hi-tech equipment in the modern industry has created awareness among the executives that system availability, reliability and maintainability should be given first importance to keep the down time of the equipment to bare minimum. High availability of equipment is dependent upon the inherent reliability and maintainability, incorporated during design. But subsequent availability is dependent upon maintenance of the equipment. It is also seen that reliability and maintainability are inherent part of functional design of equipment and their improvement can be quite difficult, costly and complex task. Maintenance, thus, should not be considered only as a simple activity to repair failed equipment but to enhance the life cycle without sacrificing the safety aspects. With this in mind, we must examine the alternative strategies in maintenance with their advantages/disadvantages, associated approaches and techniques involved to apply the same to improve the overall performance of the equipment.

TYPES OF MAINTENANCE SYSTEM

Basically maintenance system are of two types i.e. unplanned and planned. Under the former, the equipment is allowed to run until it fails or stops working. Here, maintenance is carried out in an unplanned manner, after the failure of the machines. In the later case, the maintenance is carried out as per predetermined plan.

Unplanned Maintenance Systems

Under this category, the equipment/system is allowed to operate until it stops working, which requires repair work to restart it. It can be said as corrective maintenance, which can be defined as (B3811: 1984).

The maintenance carried out after a failure has occurred and is intended to restore an item to a state in which it can perform its required function.

The failures under this category may be sudden, without prior indication or may be due to degradation of parts/components. Normally, no corrective measures are taken to avoid such failures till the equipment fails by itself. In other words, the maintenance work is required to be carried out in case of an emergency failure. When necessary repairs are carried out to bring back the equipment to its original working conditions.

From the time machines have came into existence, this type of maintenance is still in practice and is being followed in many organizations, where unscheduled stoppages cause minimal inconvenience and its advantage lies in the fact that under these conditions, it may well be a low cost option. Indeed, where, components fail randomly and provide no prior indication of impending failure, this type of maintenance is most suitable. If this system is only followed, will lead to poor operational availability of equipment, as the spare parts may not be available readily. Even though it may appear to be an economical proposition, work would greatly suffer if the equipment were not restored to operational condition immediately. Here, during repair time, no care is taken to know the real cause of failure of the breakdown, which can be helpful for future course of action.

With high-cost production systems being put in use, the limitations of the unplanned maintenance were realized and consequently other types of maintenance systems were introduced. The most important of all was planned preventive maintenance at that time. Now even, much better maintenance systems such as condition base maintenance are also being practiced.

Planned Maintenance

The planned maintenance is an organized type of maintenance, which takes care of other aspects, such as control and recording of required information for this type of work.

Under this type of maintenance, the work is planned before hand to avoid sudden failure as well as to prolong the effective operation, availability or useful life of a system, by carrying out regular inspection/repair on the equipment. It decides not only when and what type of maintenance work, but also who would undertake it? To meet the requirements of the planned maintenance, a study has to be carried out to decide the periodicity of maintenance work. Time study can also help in devising optimal maintenance schedules for the given system. Some factors that must be taken care of are as follows:

(*i*) *Utilization of equipment/service:* Here emphasis has to be on optimal utilization of machine/facility i.e. under utilization or over utilization must be avoided.

(*ii*) *Working conditions:* Attention should be given to the environmental factors such as humidity, temperature and corrosiveness which may adversely affect the performance of the equipment/facility during their use.

(*iii*) *Special factors:* Utilization of raw material for a particular environmental conditions cited above must be given due weightage to minimize breakdowns. In planned maintenance work, instructions have to be in greater detail and specific for each type of equipment. Where safety is of prime importance, the equipment must be checked regularly. Hence, the type of maintenance jobs will depend upon the nature of equipment and its working environment. Planned maintenance, therefore, involves the following type of jobs that are equipment specific.

Preventive Maintenance

The formal definition of preventive maintenance (BS 3811.1984):

Is the maintenance carried out at predetermined intervals or corresponding to prescribed criteria and intended to reduce the probability of failure or the performance degradation of an item.

One of the main objectives of the preventive maintenance is to detect any condition that may cause machine failure before such breakdown occurs. This makes it possible to plan and schedule the maintenance work without interruption in production schedule and thus improve operational availability of equipment. Under this type of maintenance, a systematic and extensive inspection of each item of equipment (or the critical parts) is made at predetermined time intervals.

It is important to note that application of preventive maintenance on all types of equipment will not be economically viable. Therefore, attempt should be made to optimize it in a cost effective sense by identifying, first the critical machinery and secondly the most efficient method to determine the periodicity of the maintenance functions. The two traditional approaches are, which include scheduled maintenance and condition base maintenance. However, in recent years some more new approaches have been evolved, such as reliability centered maintenance (RCM) and total productive maintenance.

Scheduled Maintenance

In this type of maintenance work, the total maintenance program is scheduled in consultation with the production department so that the particular equipment is made available for maintenance work. For this purpose, accurate failure data must be made available to establish machine failure patterns and therefore, maintenance interval. This program has some difficulties/demerits and should be examined before its application to any system. The main difficulties associated with this type of system are the following:

(a) *Determination of Time Interval*
   (i) Where some of the components/parts may be changed without having failed due to risk, which will add to extra maintenance cost.
   (ii) At the same time, some parts are not replaced as and when needed, may add inconvenience and cost during their actual failure.

(b) *Environmental Impact:* For the calculation of time interval i.e. failure frequency, the impact of actual working environment is difficult to be considered, which may make considerable difference. However, this aspect must be given due weightage for the assessment of components/parts life.

(c) *Availability of Failure data:* In most of the cases, collection of authentic and accurate data is difficult and therefore, the estimates are based on the probability theory, which may be away from the actual conditions of the equipment.

As the frequency of such maintenance work is predetermined, it is possible to utilize the idle item of the equipment to carry out maintenance work. This work also helps, the maintenance department to make effective use of their available manpower. The help of experts in the field can also be taken, since the nature of work is known in advance. Though the scheduled maintenance is costly as compared with breakdown maintenance but the equipment availability can be improved with safety.

Condition Based Maintenance

Formally condition based maintenance is defined as (BS 3811.1984).

The preventive maintenance initiated as a result of knowledge of the condition of an item from routine or continuous monitoring.

This policy is most suited for high capital cost equipment and complex replaceable items and also for permanent parts that deteriorate over time or due to prolong use. For this purpose good knowledge of failure data is essential for effective implementation of condition based maintenance. This process involves direction of diagnosis and proper execution of maintenance planning on indication of deterioration. The unit/system can be shut down at a convenient time to avoid accidental breakdown. The advantages of the condition-based maintenance include the following:

(*i*) Detection of impending failure, through, the use of trend analysis or qualitative techniques, which will reduce the failure rates and consequently improves the availability and the reliability of the equipment under use.

(*ii*) With the information of failure rate, it is possible to shut down the plant at a convenient time to carry out maintenance work without interrupting the production.

(*iii*) With condition-based maintenance, it is possible to pinpoint the impending failure with its cause, which may allow some modification, if needed. The system maintainability can also be improved with better spare parts management.

(*iv*) The most important advantage is the flexibility of the system, if used with adoptive control, to enhance the useful life of the equipment, by limiting the dynamic failures within the system and thus improves the system reliability.

To achieve an accurate failure prediction, an effective monitoring method, to know the status of an item or component is essential. This can be done with the help of condition monitoring (CM), which is of two types:

(*i*) On-line monitoring, where machinery can be monitored while it is operative.

(*ii*) Off-line monitoring, in which case the system has to be shut down for check up.

The use of condition monitoring helps in formulating all necessary maintenance actions, to be undertaken at an appropriate time to the predicted failure and ultimately enhances the system availability and reliability. Decisions in this case are based on the diagnostic information, thereby providing cost effective measures as well. The benefits derived from condition monitoring include technical, financial as well as organizational and attitudinal, which help in organizational growth.

Reliability Centered Maintenance (RCM)

It is used to identify maintenance requirements of machinery. The RCM establishes the functional requirements and the desired performance standards and equipment and then these are related to design and inherent reliability parameters of the machine. For each function, the associated functional failure is defined, and the failure modes and effects of the function failures are analyzed.

The consequences of each failure are established, which fall in one of the four categories: hidden, safety or environmental, operational and non-operational. Following the RCM logic, pre-emptive maintenance tasks, which will prevent these consequences, are selected, and provided the applicability and effectiveness criteria for preventive maintenance are satisfied.

The applicability requirements refer to the technical characteristics for possible maintenance tasks and the frequency at which these should be carried out. Effectiveness criteria depend on the consequences of failures, probabilities of the multiple failures for hidden failure consequences, acceptable low risk of failure for safety consequences and non-operational consequences. When the requirements for planned maintenance are not fulfilled, default tests include failure finding (for hidden failure, possible redesign of equipment procedures and training process) for no scheduled maintenance.

This was first applied to critical problematic areas to reduce the maintenance cost in a food processing plant. RCM helped in reducing the downtime of the equipment besides achieving reduction in scheduled maintenance load and effecting improvements in system/equipment availability. With the development of technology RCM can be applied to all types of costly and complex equipment/system.

Total Productive Maintenance (TPM)

This approach to maintenance was evolved in Japan as a development of preventive maintenance and after passing through various stages, finally became total productive maintenance. The various development stages include, incorporation and application of breakdown maintenance, reliability and maintainability theory with concern for economic efficiency in plant design. This was made possible with a comprehensive maintenance system based on total employee participation. The user of the equipment carries out this maintenance with all available methods within the user company.

In this case the maintenance is based on the situations within the organization and the policy of maintenance may differ from machine to machine to meet out cost effective criteria. Under TPM the conflicts of production and maintenance group are minimized since everybody is responsible of all the activities of the organization. Improvements in this method are possible to achieve a zero equipment failure situation and hence, maximum equipment availability, productivity and quality of production. It also eliminates the independency on out side expertise in equipment maintenance. The objective of TPM is to prevent maintenance, improve maintainability and practice preventive maintenance for the complete life cycle of the equipment.

MAINTENANCE PLANNING AND SCHEDULING

To improve the operational availability of the equipment/system, the maintenance activities must be planned effectively. The first step is to establish the number of machines, which are to be maintained. Next, it is to be decided that how these equipment/facilities will be maintained. Then a maintenance schedule is prepared for all critical items, which require preventive maintenance. Initially, this exercise can be started with a few selected components of machines and with the experience gained; it can be extended to other parts of machine under consideration.

The main purpose of maintenance planning and scheduling is to raise the maintenance standards and improve cost effectiveness. This is only possible through critical analysis of the results of maintenance. In case, where number of equipment/machine is very large, it is difficult to prepare maintenance schedules manually and therefore the use of computer becomes essential. When the planned maintenance works are not computed in time, rescheduling of the work is done to clear the backlog jobs.

Once the process of schedule generation is completed all assigned works must progress smoothly until something comes up that necessitates alteration in the predefined process.

Sometimes due to budgetary constraints, rescheduling of maintenance jobs becomes necessary. It is then required to work out schedules in such a way that backlogging is avoided. This is mostly desired when a fixed scheduling process is followed. In the case of dynamic scheduling, the system takes care of the work done in the past and that to be done in the future. The fixed scheduling works well in an ideal environment, however, dynamic scheduling takes care of all practical constraints under all circumstances.

For proper scheduling of the maintenance work, the jobs must be controlled through a work order system, which provides the basic paper work for planning the work force, material and time. A work order is also an authorization to carry out the job.

Priorities must be established to handle a mixture of backlogged and new pieces of scheduled jobs. The frequency of scheduled maintenance and the criticality of operations are the most meaningful attributes for establishing a priority for a job. Frequency indicates the relative importance of one procedure over the other. If equipment becomes overdue for certain number of works, it is treated critically important and a critical point is said to have reached. To minimize the critical points, more emphasis should be placed on scheduling the work, when it is due. The jobs should be scheduled according to priority system followed in a particular industry.

MAINTENANCE EVALUATION

The performance of the maintenance department may become pivotal in assuring the economic competence of any project since maintenance cost is increasing day by day due to sophistication of equipment and degree of accuracy required. It is of prime importance to employ effective measures for maintenance of jobs to improve equipment availability and achieve reduction in maintenance cost. The success of any organization depends on the effective utilization of its manpower and resources. The most important consideration in this regard is to identify the parameters that will measure the maintenance performance. These parameters will be used as maintenance evaluation index (MEI) and be of much value as tools of any decision support system for project management.

Comparing the results obtained with and without the work measurement programs, can recognize the need of maintenance evaluation. By work sampling, the cause of poor performance and the possibility of improvement can both be identified. Due to involvement of more maintenance personnel in modern organization, the need of maintenance evaluation has further increased. The main objectives of maintenance evaluation are:

- Avoidance of overstaffing and understaffing
- Setting up, the standards for different types of jobs in advance
- Evaluation of maintenance functions, through indices.

The evaluation will also indicate the level of maintenance, existing in various organizations. This will give a fair idea about the overall performance of the organization. The performance analysis at different levels, starting from foreman to the maintenance engineer can also improve the work culture. This will give the organizations/departments a clear picture, where they stand in the fast changing technical environment in order to be able to carry out effective improvement of the maintenance function. However, maintenance functions are affected by numerous variables. Therefore, it is experimentally difficult and complex, if not possible to measure the effectiveness of maintenance in absolute terms. Poor housekeeping, shabby appearance, and lower machine life are all indicative of bad maintenance.

The benefits that are achieved through proper evaluation of maintenance include, performance improvement, reduction in delay, reduction in equipment downtime and improvement in preventive maintenance.

The evaluation of maintenance function can be carried out by comparison with the similar type of work. If information about the same type of work is not available from elsewhere, then the evaluation of maintenance work can be performed over a period of time. For effective evaluation of the maintenance function, it is desirable to carry it out at different levels with the maintenance department. A three-tier-evaluation system is possible, the first being at foreman level, which can indicate utilization of maintenance crew and the time spent for maintenance/ repair equipment wise.

The second and most important level of maintenance evaluation is at the level of deputy manager, who is responsible for guiding the foreman in matters of control of manpower and the cost involved in maintenance work. The important parameter for performance evaluation may include labour productivity, availability of machines time, cost, materials and spare parts needed to do the job.

The last level of evaluation is at the level of the maintenance manager. The indices developed, help the maintenance manager in evaluating the performance of the maintenance department as a whole and enable him to draw up plans for improvement of the maintenance function.

The types of maintenance evaluation techniques followed in practice include the following:

- Evaluation through reports
- Subjective evaluation
- Objective evaluation
- Work auditing

After evaluation, statistical analysis of performance is carried out for fixing the indices.

MAINTENANCE CONTROL

Maintenance is the requirement of any machine and system to keep it in working order. It also involves cost therefore its control is very much essential. There are number of maintenance policies available and each one of them will not suit for all types of equipment/machinery. Before adopting any type of maintenance policy techno-economical analysis must be carried out to establish its feasibility. The designers and manufacturers do provide certain guidelines for their products, but it is the duty of the user to select a suitable maintenance policy looking into their need and environmental conditions.

It is the normal practice that during major overhauls, certain parts/components need replacement as recommended by the manufacturer. But during preventive maintenance/ breakdown maintenance some of these parts/components must also have been replaced without completing their stipulated life. Due care has to be taken to avoid duplication of such components to keep the maintenance cost under control.

The other area of control is the spare parts management which must be given due attention to control the maintenance cost. Some spare parts have a particular life cycle and therefore must be used accordingly. After expiry date their use is not permitted. The same conditions apply to the lubricants, which are used in almost all machines/equipment. During operating life cycle of equipment replacement of lube oil is carried out regularly. But again, at the time of scheduled maintenance, it may be replaced completely, which means that some fresh oil is lost.

Availability of spare parts during the course of maintenance is essential and if not provided in time will add to the maintenance cost. The control of manpower for maintenance function needs due attention since the overtime cost is always higher than the normal wages. To carry out the maintenance functions in a pre-specified budget, the following measures must be adopted.

- First line supervisors must be apprised of the cost information of the various materials so that the objectives of the organization can be met without extra expenditure on maintenance functions.
- A monthly review of the budget provisions and expenditures actually incurred in respect of each center/shop will provide guidelines to the departmental head to exercise better control.
- Total expenditure can be spread over the year uniformly to avoid over spending and under spending at the end of the year.
- The concept of total productive maintenance must be followed as far as practicable, where participation of all the employees for all the jobs of the organization is considered, which will ultimately exercise control by itself.

A timely submission of equipment/machinery for service/repair by the production department can contribute significantly towards reducing the rate of material consumption because it would avoid further deterioration of parts, which can be repaired and therefore, avoid replacement of parts/components.

If the average life cycle cost of new equipment is known and the combined cost of operating and maintaining an old machine is evaluated, the exact time for replacement can be decided and saving of some expenses can be obtained.

Efficient planning of maintenance functions is one of the most important ways to minimize the maintenance costs. Planning ensures that material, labour; tools, work place, etc. are properly arranged and utilized. This helps the maintenance crew to complete the job in a well-organized and productive manner. It is seen in practice that a good amount of time, is wasted by maintenance crew in laying their hands on the above facilities and such in fructuous efforts can be minimized through better planning and control. The supervision of maintenance is also essential to control the costs.

MOTIVATION IN MAINTENANCE

In a number of Indian production systems, the availability is poor. This is due to lack of adequate steps to improve the maintenance organization. One important cause of poor maintenance performance is the low motivation to work. Maintenance staff can be motivated through proper training, through providing proper incentives, through labour performance analysis and control and through maintenance based compensation. The concept of overtime to improve productivity has not yielded the desired results. Motivation means stimulation to act. Stimulation may differ from one type of job to another. The motivational efforts should be directed at stimulating men to work for achieving the desired goal of the organization. However, the true motives of individuals are not always apparent and are best known to the individuals themselves. The actual motives of individuals may be to achieve immediate goals such as money, security, or prestige. Adequate importance therefore, should be given to any type of job related to maintenance to keep the moral high, of the maintenance workforce. Following are the activities, which can motivate the maintenance personnel:

Maintenance Training

With the advancement of technology, automation, electronic controls and sophisticated mechanisms have been introduced. To maintain such equipment, it is essential to train and develop personnel for this type of work. Today's maintenance engineer needs a totally new set of skills and considerable knowledge in a variety of complex field such as electronics, instrumentation, control systems, data acquisition system, computers, non destructive testing, robotics, etc. Training is a continuous process and needs to be adequately planned to update the knowledge of all working personnel. This program can be in the form of apprentice training, concurrent training, long-term training and short-term training depending upon the need of the organization.

Incentives in Maintenance

As practiced in production departments, financial incentives can be given to maintenance personnel to motivate them. Overtime can be one form of incentive. Some rewards for better maintenance work should also be instituted by the organization as an incentive plan, but this must be based on the performance and quality of work carried out.

Labour Performance Analysis

It is easier for the maintenance engineer to control the material and overhead costs than the labour costs. Therefore, it is important that cost analysis be carried out in respect of maintenance work force, in the form of man-hour utilization or work output to improve the labour performance. The performance of labour force must be evaluated individually or group-wise and the same may be displayed for the information of all concerned. The best worker must be appreciated and the worst must be punished.

Maintenance Based Compensation

The work of maintenance must be treated at par with production department, which was not done earlier. The concept of contract can also be applied to maintenance work if it is of non-repetitive in nature. This helps the management to reduce the downtime of the equipment. If the work is completed before stipulated time period, incentives can be given, whereas for delays, penalties can be imposed. For the very reasons, the contract maintenance is becoming common in the areas of computers, air conditioning, photocopiers, etc. The basic advantage of contract maintenance is to get the services of experts in the respective areas.

MAINTAINABILITY

It is characteristics of a product design and installation which is expressed in terms of ease and economy of maintenance functions. The objective of maintainability is to design and develop a system which can be maintained in the least possible time at minimum cost. This process starts from design development, production, installation and testing of equipment/system. It consists of four stages i.e. planning, design application, measurement and evaluation to meet specific operational needs. The success of this program contributes directly to the profits from sales. Maintainability is one important component of design team to provide with the specialized information and data for consideration in the definition of a system.

The concept of maintainability started in military services of US in 1954 since it was difficult to obtain 100 percent reliability of equipment/system. However, the reliability programs were aimed to improve the service life of the products. With the development in technology the cost of maintenance also increased appreciably, which forced the design engineers to make products which were easy to maintain at low cost. In early 1966, attempts were made to standardize the approach for the implementation of maintainability program, its techniques, and procedures to demonstrate it. Later on it was associated with the measurements of cost effectiveness, system effectiveness and integrated logistics support.

To understand maintainability the terms like maintenance, preventive/corrective, MTBM, MTBR, availability, cost effectiveness etc. must be clearly defined. The same has been discussed at appropriate places. Maintainability affects availability of equipment/system directly, as the time taken to repair failed parts makes it unavailable for use.

Maintainability Concept

When the product is ready for use, it is said to be available. The availability is the ability of product to perform its pre defined functions when required in use and depends on the occurrence of failures, and time required for repairing it. The system is either in up state (available) or down state (not available). The up state includes standby period and active use time of the equipment whereas, the down time includes the active repair time and waiting time for spare parts etc. The availability calculations have been discussed in detail in earlier chapters.

Maintainability Programs

The maintainability can be programmed in the same way as product performance or reliability and consist of the following stages:

(*i*) **Program initiation:** It starts from the definition of all tasks required for maintainability program monitoring and controlling till its execution,

(*ii*) **Design and analysis:** During product design stage goals for maintainability are set, analysis and predictions are made.

(*iii*) **Evaluation and test:** With the evaluation and test of the maintainability program, it is demonstrated that the goals set have been achieved successfully.

Maintainability Modeling

Depending upon the complexity of equipment or its importance, a formal model for maintainability can be generated in a quantitative form. Model here means to establish maintainability predictions.

Maintainability Design

The following are the maintainability criteria identified after knowing its requirements:

- Identification and incorporation of all possible built in self-testing capabilities.
- Monitoring of equipment without the use of extra test equipment or special tools.
- Identification and incorporation of fault isolation points to minimize repair time.
- Attachment of major assemblies is made directly assessable and interchangeable without requirement of maintenance alignment or adjustments.

To help in establishing various maintenance/repair policies the following questions may be employed as a guide to ensure complete coverage.

- What are the quantitative maintenance/maintainability requirements at various levels of maintenance?
- What are the anticipated scheduled and unscheduled maintenance functions for each level of maintenance?
- What are the manpower requirements with the skills for each level of maintenance?
- What are the basic constraints of maintenance support system?
- What are the facility requirements of each level of maintenance?
- What is the maintenance environment associated with each level of maintenance?

Maintainability Analysis

This constitutes the process of translating system level operational and maintenance requirements into specific qualitative and quantitative equipment design requirements. The analysis provides:

(*i*) A comprehensive documentation of specific maintainability requirements to be met during design phase.
(*ii*) A list of check points to ensure that all features necessary to meet above requirements have been incorporated.
(*iii*) A summary of basic maintenance functions and support requirements specified for system.

Analysis process starts from functional analysis which assures:

(*i*) That all elements of the product are fully recognized and defined.
(*ii*) A measure of relating equipment packaging concepts and support requirements to given functions.
(*iii*) A common reference point for all subsequent engineering system design and development effort.

Maintainability Allocation

When the maintainability program plan has been prepared and functional analysis and maintenance concept data have been updated, then design and performance specifications for the system are prepared. The specifications become the primary means whereby maintainability requirements operate as constraints on the system designer. Accordingly the maintainability engineer must ensure that relevant maintainability requirements have been incorporated in every specification. This is done by maintainability allocation.

An allocation consists of determining the contribution of active down time of each component and evaluation of the contributions of all such components against established mean time to repair for each system.

Maintainability Prediction

The need for maintainability prediction aims at, to find the expected number of hours that a system/equipment will remain inoperative or in down state while it is under maintenance. The specific advantage of using maintainability prediction procedure is that it highlights areas of poor maintainability to the design engineer which justify product improvement, modification or change in design.

A good prediction technique should be used to directly pin-point design deficient or support deficient areas for improvement. The elements of prediction are (*a*) failure rates of components/parts and (b) repair time required at the maintenance level. The maintainability prediction procedure includes (*i*) applicability (*ii*) point of application (*iii*) basic parameters of measure (*iv*) information required and (*v*) cautionary notes.

After collecting the above information the equipment/system maintainability can be made by using the available techniques.

TOTAL QUALITY MANAGEMENT

It has been discussed earlier that quality is directly associated with reliability of the product/ system. During the last three decades organizations have laid emphasis on the quality of their products to compete in a global market. With the inception of globalization and creation of international markets have led to standardization of quality system and the concept of total quality management (TQM) has gained momentum. The term quality has undergone conceptual change with process of evaluation. Initially it was associated with inspection and later on covered the areas of process control, quality assurance, total quality management and strategic quality management.

In the initial stage quality was confined to shop floor activities by means of inspection and process control. The concept of quality assurance took care of design process. With active participation of working personnel the concept of 'Total Quality Management' was evolved. Further, the concept of 'Strategic Quality Management' (SQM) was developed with active cooperation of customers to maintain quality of the product throughout its life. It is well understood by the industries that the quality is the most critical factor for their survival in the open market. Improved quality is often advantageous even if it does not increase the market share. There is evidence to suggest that customers prefer better quality at a higher price for achieving trouble free service from it. Therefore, it is advisable to produce less number of items of quality rather than producing more and increasing the number of defective items.

TQM is an active approach for providing continuous improvement of quality with the cooperation of every one involved. It was a thought of two US citizens namely W.E. Deming and J.M. Juran who launched it in Japan after World War II.

In real sense the quality means a predictable degree of uniformity and adoptability at a low cost suited to the market. Various researchers have given definitions of quality almost with similar thought of improving the product/system.

Characteristics of Quality

The basic characteristics of quality are the following:

1. Quality depends upon evaluation made by persons who are not involved in production.

2. **Features:** For example a TV have a remote control.
3. **Reliability:** The equipment will render trouble free service for a specified time period.
4. **Serviceability:** How easy or difficult is to carry out maintenance as and when required.
5. **Durability:** What will be the life of product.
6. **Conformance:** It measures how the products/equipment will meet the specifications or target set by the designer.
7. **Aesthetic characteristics:** How an item/product look, feels, tastes, or smells. These are more subjective and difficult to measure.
8. **Perceived quality:** Feeling of confidence in the level of quality felt by the customer.

Cost of Quality

The cost of quality comprises of the following two parameters:

- Cost to control (prevention and appraisal)
- Cost of failure of quality (internal or external failures)

When the product does not confirm with the given specifications it is the cost of the organization whereas, on success of completion of the task, it is the asset of the company. However, the quality cost can be divided into four groups as discussed below:

(*i*) **Prevention cost:** This cost is incurred to prevent production of defective parts/ components by adding design reviews, quality orientation programs, process control, workers training, etc.

(*ii*) **Appraisal cost:** This cost is incurred in auditing the production work to meet out design specifications by means of process capability measurement, testing equipment, prototype inspection, in-process and final inspection work in-process testing and inspection.

(*iii*) **Internal failure cost:** This cost includes the expenses made for production, the product and not in position to sell it in time due to some reasons of redesign, reprocess, retesting etc.

(*iv*) **External failure cost:** These costs are associated when the product is manufactured but some money is lost due to warranty costs, product liability field services and product return etc. The following figures show the variation of quality cost with passage of time.

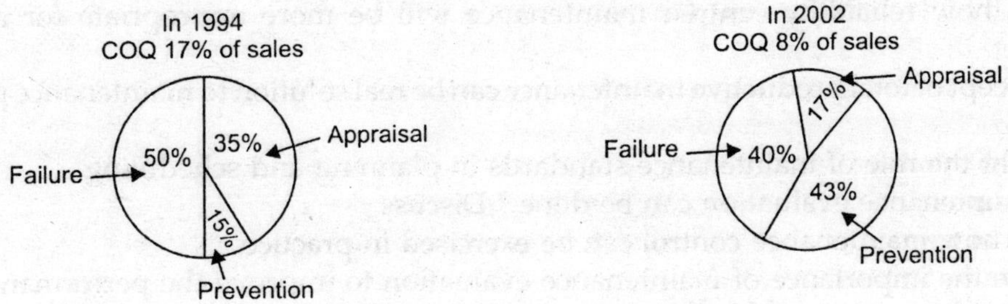

Fig. 13.1: Impact of cost of quality (COQ)

EVALUATION OF QUALITY MANAGEMENT

The growth of quality management is very rapid and in recent past simple inspection activities have been supplemented by quality control. Quality assurance has been developed and redefined as inspection, quality control, quality assurance and total quality management.

CONSEQUENCES OF TOTAL QUALITY

The basic objectives of any productive/service system are customer satisfaction, compliance of specification, market share, higher productivity, zero defects, increase in sale, decrease in cost, minimization of down time, scrap and rework.

The above objectives will become mere consequences if total quality system is followed. It is evident that the cost of quality control can be compensated by customer satisfaction, profits, market share, improved work morale, innovation etc. Further it will ensure TQM environment with improvement in overall performance of any organization.

For the purpose of improving quality of product valuable tools of quality such as Taguchi, Deming can be applied. Further use of quality circles can also be made. The certification of it can be obtained through ISO 9000.

QUALITY AUDIT

The level and nature of audit can be done with the help of ISO 10011 -1, 2, 3. Part - I deals with auditing supported by Pt - II - guidelines for auditing quality systems and Pt - III - management of audit programs. The standard advises that each organization should develop its own specific procedures for implementing the guidelines. The audit however, should be carried out by outside agency which is not involved in production. The auditor should have requisite qualification for the work.

QUESTIONS

1. Discuss the various maintenance systems being practiced.
2. Differentiate between planned and unplanned maintenance systems.
3. Highlight the importance of preventive maintenance and how it differs from scheduled maintenance?
4. What is condition based maintenance ? Elaborate.
5. Discuss how reliability centred maintenance will be more appropriate for automatic plants.
6. The concept of total productive maintenance can be real solution to maintenance problems. Explain.
7. Highlight the role of maintenance standards in planning and scheduling.
8. How maintenance evaluation can be done.? Discuss.
9. Discuss how maintenance control can be exercised in practice.
10. Illustrate the importance of maintenance evaluation to improve the performance.
11. How incentives can improve maintenance personnel performance? Explain.
12. What is the role of maintainability in maintenance functions? Explain it briefly.
13. What is total quality management and how it is helpful in maintenance system?

14

Condition Based Maintenance

INTRODUCTION

A high level of machinery/system's, availability is essential in all automated and capital-intensive equipment and organizations. Unplanned shutdowns are the major problem for maintenance executives. Modern and improved maintenance techniques can minimize random failures as well as, can provide higher equipment availability and reliability.

Condition based maintenance (CBM) is one, out of a number of systematic approaches which can be used under planned maintenance. The philosophy of CBM is primarily based on terro-technology. The approach requires the use and analysis of actual condition data, to suggest the type of maintenance function to be adopted for particular equipment. In the service period of equipment, the availability of equipment/system depends upon maintenance policy being adopted. The availability can be expressed in terms of mean time between failure (MTBF) and mean time to repair (MTTR) of the equipment, after its failure.

$$\text{Availability} = \frac{\text{MTBF}}{\text{MTBF} + \text{MTTR}} * 100 \qquad \text{...(14.1)}$$

It is obvious from the above that increase in MTBF and decrease in MTTR will affect the system availability. This is also related with the system reliability and maintainability.

Condition based maintenance leads to undertaking corrective measures only when they are needed rather than at the scheduled or routine intervals. Thus, it can eliminate downtime for unnecessary inspections. Most of the processing and power industries utilize some form of condition-based maintenance for higher reliability of their equipment/system. Normally regular inspections are carried out at a specified interval to examine the condition of parts/components by application of fault diagnosis techniques. There are number of such techniques available to identify the causes of impending failures in machinery.

The conditions of equipment/machinery are compared against acceptable pre-set standards, which have established before hand during commissioning tests. Condition based maintenance is considered as the most reliable, cost effective and efficient technique for maintaining majority of critical equipment such as engines, turbines, compressors, etc. used in most of the industries. It is observed through studies that 20-30 per cent of possible production times go towards maintenance of equipment/machinery and this can be minimized through effective maintenance. Towards this aim, CBM plays a very important role. Thus, it is necessary to develop an appropriate monitoring system in the overall maintenance schemes for all machinery.

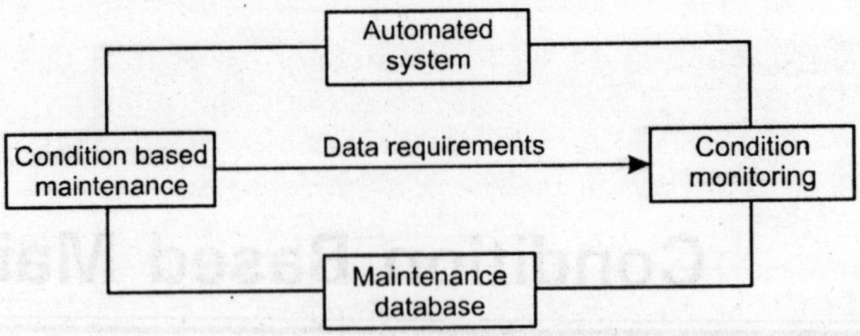

Fig.14.1: Condition Monitoring in Computer based Maintenance

It is seen from Fig. 14.1 that by using condition based maintenance in any automated manufacturing/processing organization, sufficient data can be obtained which can be applied for condition monitoring of the equipment/system. Through condition monitoring also, maintenance database can be generated, which further can be utilized for condition-based maintenance planning and control.

Therefore, condition monitoring is the means by which the condition-based maintenance can be carried out through various available techniques to identify the impending failures. The use of condition monitoring ensures that all possible maintenance functions can be done at a appropriate time to the predicted failure and therefore, improves the equipment availability through organized maintenance actions. Under condition monitoring a periodic or continuous comparative measurement on parameters, which can indicate the condition of component/part or system is carried out to assess the actual condition with pre-set values of measurement taken during the installation of equipment/machinery. This also helps in planning of future maintenance action plans in advance, therefore, providing a cost effective and logical decision-making by the concerned executives.

CONDITION MONITORING

As indicated above, the necessity of condition monitoring arises, where the cost of safety-driven, failure preventive servicing programs is required to improve the equipment/system availability. This also contributes to maintenance planning; maintenance cost reduction, health and safety programs and helps in energy conservation too, by providing early warning of waste and inefficiency arising from faulty operation. The work of maintenance planning is facilitated by advance appraisal of defects through condition monitoring. Such pre-planning also helps to extend support to several aspects of maintenance work such as, spare parts, technical information, and manpower requirements. It also helps in completing the maintenance work in scheduled time, thereby reducing overtime payments and consequently the maintenance costs.

Monitoring techniques not only enable equipment to operate at optimum performance with improved reliability but also provide trends in machine deterioration to serve the purpose of scheduling constructive maintenance without affecting production.

Levels of Condition Monitoring

Condition monitoring can be carried out at different four levels. First level is visual inspection by seeing, hearing, touching and smelling. Second level through sensor assisted inspection by using various instruments such as, tachometers, odometer, test meters, proxi meters, pyrometers, corrosion meters, and vibration and frequency analyzers. Condition of lubricants and engine oil etc can be monitored through advanced techniques at level three, by using indicator analysis. This indication may be in the form of metal particles present in the oil. The quantity of which will indirectly indicate the condition of the system under operation.

The level four monitoring can be carried out with the help of mini-computers or microprocessors. Here, process transducers, accelerometers, counters and other sensors are employed at different points of the machinery for acquisition of data to indicate failure of the system. The use of data loggers and computer networking and data transfer has all revolutionized at the level four condition monitoring systems. However, at level four condition monitoring requires costly monitoring gazettes and therefore, suitable for safety oriented and high value equipment/machinery.

CONDITION MONITORING TECHNIQUES

The success of condition-based maintenance depends on the efficiency of identifying the deteriorating trend in the machine components. For this purpose, it is essential to recognize the sources or causes of failure. Failures of different types induce different types of effects on the system. These effects may be classified as dynamic effect, particle contaminant effect, chemical effect, physical effect, temperature effect and corrosion effect. The condition monitoring system aims at, recognizing these effects and assessing their degree with the help of some specially, designed devices. These devices/instruments provide a quantitative measurement of the effects and thus help to assess the condition of machine/equipment. Some of these are briefly discussed below:

Dynamic Effects

These effects can be identified through vibration analysis, pulse monitoring and acoustic emission analysis. The vibration signals carry information about exciting forces and the structural path through which they propagate to vibration transducers. A machine generates vibration of specific pattern, when it is in healthy state and the pattern of it changes during service life of the machine/equipment due to degradation of component/parts. Thus vibration monitoring, is used to monitor the condition of the machine/system. There are few advantages of condition monitoring which include, cost reduction in maintenance, less number of failure, damage to components/parts and confidence gained by the operational personnel.

Vibration measurement can be made with the help of transducer, vibration pick-ups or strain gauges. Mechanical vibration can be measured in terms of displacement, velocity or acceleration. Low frequency vibration is best measured using displacement, and as the frequency increases use of velocity and then acceleration becomes more suitable. However, the use of Piezo-Electric Accelerometer is most suitable for wide range of frequency, because it has good linearity throughout its wide frequency range. For low speed machines velocity transducers are preferred. Proximity sensors may be used to measure shaft eccentricity or misalignment.

The Vibration Analyzer

It is used, to pick up the signals from the working system in order to study the condition of component/machine. With the help of vibration analyzer, the complex waveforms can be analysed by means of narrow band filters. The frequency range obtained, enables the maintenance engineers to prepare routine check ups on rotating machines/parts. In complex machineries, the individual component vibration frequencies as well as, overall vibration level can be monitored.

The Piezo-Electric Accelerometer

This is general purpose and widely used vibration sensor, which uses a piezo-electric ceramic or pure crystal element sandwiched between a mass and casing. The sensor is so designed that it exerts either a compressive or shear force on the element. Its design does not involve moving parts and therefore it is more reliable, rugged and not subjected to wear besides its lightweight. Special type instruments are available for application in three-dimensional planes, high temperature environments, very low vibration level and high-level shocks. Small types are used on delicate structures such as printed circuits boards and for very high frequency measurements. Mounting of accelerometer affects the accuracy of the vibration measurements and therefore, due care should be taken. Its location on the monitored system is also critical, for the success of monitoring therefore, it must be located close to the monitored component to maximize sensitivity. When used on rotating parts, due care should be taken for its monitoring because its position i.e. axial or radial will give different results.

Shock Pulse Meter (SPM)

These are used to monitor oil film thickness in a bearing and the conditions to support and maintain this film, as well as bearing disorder. Under vibration spectrum, it is difficult to detect a defective bearing because its contribution to the spectrum is very small. Whereas SPM measures the mechanical impacts generated by a rolling element bearing. These impacts generate pressure waves that are transmitted through bearing housing.

Fluid Condition and Particle Monitoring

By monitoring fluid, it is possible to detect the presence of solid particles or other contaminants, which can lead to the failure of any system where lubrication oils and hydraulic fluids are used. The change of viscosity of fluid due to degradation can also lead to system failures and reduction in useful life cycles. The physical condition of fluid can be measured in terms of viscosity, water content, acidity and basicity.

The lubricating oils are used to provide fluid films between two mating parts and change in the viscosity of lube oil affects the film formation, which results in temperature rise, oxidation and increased wear rates. Increase in acidity causes corrosion of the components/parts where hydraulic oil is used. The main contaminants that could be present in the lubricants/hydraulic oil are aluminum, iron, antimony, lead, silicon, chromate, silver copper, tin, etc

To examine the presence of foreign materials in lube oils, spectrometer oil analysis procedure (SOAP) and magnetic plug inspection system, besides Ferro graph and X-ray spectrometry can be used.

Spectrometry

This is a well-established and widely used method of fluid analysis, which can provide data on the composition of debris. It can identify metals and can also detect lubricant additives of very small size i.e. 5 to 8 microns. The techniques used include atomic absorption, atomic emission and X-ray fluorescence.

In atomic absorption spectrometry, a sample of fluid is diluted and vapourized in an acetylene flame and compared with reference standard for direct measurement. The atomic emissions spectrometer, a high direct voltage (1500V) excitation of the metallic impurities in the oil sample causes different types of radiation depending upon the quality of oil, which can be spectrally analyzed and compared with reference spectra. It can detect bigger particles. In case of X-ray fluorescence, the fluid is irradiated by an X-ray source to change the energy level of atoms to indicate the concentration of foreign elements.

The inductance sensors coiled on non-metallic pipes can be used to detect ferrous and non-ferrous particles. The ferrous particles will increase the inductance whilst non-ferrous metallic particles will cause decrease in the flux and therefore, reduction in inductance. This method can only detect large size particles. In the same manner light sensors can be used to detect debris in a flowing fluid.

Chemical Effects

This can be monitored by the measurement of certain traceable quantities of chemicals that are an indication of wear; leakage and changes in lube oils. For this purpose SOAP is an effective laboratory based technique, which analyses the wear particles (10-micron) held in suspension in oil and assesses their concentrations. The basic steps, required in the analysis, involve sampling, spectrometric analysis, data interpretation, and validation of results. This method has been described earlier.

Physical Effects

Some features can be detected through physical measurements of certain quantities and are called the primary effects. These can be recognized through measurable changes in the performance of the machinery for example, any significant change in power loss can be associated with an increase in friction in the system or with leakage of hydraulic oil from the system. Various measurable quantities, which need attention, are given below:

1. *Lubricating Oil Consumption Rate:* The lube oil used for the purpose of lubrication can indicate the condition of the machinery. If the rate increases abruptly, indicates that there is a possibility of the leakage of lube oil through some parts, for example, piston rings in a cylinder.
2. *Air Pressure Ratios:* In the engines, where super chargers are used, to improve the efficiency, the air pressure ratios play a significant role and must be checked timely to maintain the engine performance at a desired level.
3. *Temperature Changes:* In the machinery where cooling is used, the temperature monitoring is very important for the life of the system, since abnormal rise may seize, which is very common in case of internal combustion engines.
4. *Vibration Change:* Detonation, knock, etc in case of internal combustion engines can cause serious vibrations, and therefore may cause premature failure of the system. This will also cause power loss of the engine, if severe vibrations are noticed in the bearings or other moving parts.

Other physical effects can be monitored through liquid dye penetrants, magnetic particle inspections, ultrasonic and by visual testing of the machinery.

Corrosion Effects

Corrosion is a regular time dependent phenomenon and entirely depends on the environmental conditions. By monitoring proper environmental conditions, this can be minimized. There are various techniques, which can be used to monitor corrosion and include linear polarization resistance method (corrator), corrosion potential measurement and weight loss methods.

As indicated earlier, the condition monitoring techniques are time consuming and costly. Therefore, their use must consider the following factors for cost effective utilization:

- Criticality of the part/components used in the complete system.
- Availability of standby unit to meet out emergency requirements.
- Standardization of items to provide easy inter-changeability of components/parts.
- Operating conditions.
- Failure statistics (MTBF, MTTR) etc.
- Cost of monitoring.
- Cost of failure/penalty, in case of equipment/system under warrantee period and service requirements.
- Cost of maintenance vs replacement cost of the equipment.

Considering the above factors, the frequency of the condition monitoring must be decided to optimise the total maintenance cost.

SYSTEM APPROACH TO CONDITION MONITORING

In a traditional approach, the components of a equipment are monitored independently using sensors. These components/parts form sub-assemblies and in turn a complete system. System engineering is widely used in the manufacturing field; therefore, an attempt is made here to consider monitoring of a equipment as a system.

Monitoring of individual component/part involves recording of information for result oriented operational control, which is not efficient. Here for monitoring, a separate sensor is to be used independently, to measure each characteristics of a component/part for example, bearings. When the concept of whole system is employed, the output of one sensor can give information, on more than one sub-assembly. Therefore, monitoring at sub-assembly level reduces the number of sensors, which is the concept of system condition monitoring technique. Classic system representation uses the Lap lace transform representation of the differential equations describing the dynamic system under consideration. Practical condition monitoring does not need the analytic tools used in dynamic analysis but the concepts of the systems approach of input/output relationship.

Condition monitoring diagnosis is important to know the state of the system/machinery, which can be done by either routine or continuous mode. Depending upon the signal received the failures can be defined as hard, when large changes from specifications are observed and soft, when shift from specifications is minute and system structure is unaltered. Signature matching can make this comparison, which is a very common mode in vibration analysis.

APPLICATION OF SYSTEM CONDITION MONITORING

The concept of system condition monitoring as stated above can be effectively applied for flexible manufacturing systems and fully automated systems used for product manufacturing.

For hard faults diagnostics, the complete system/equipment is divided into sub-assemblies depending upon the nature and frequency of failures. This can be done with the help of failure mode and effect analysis (FMEA) and fault tree analysis (FTA) to produce fault nature, which gives failure signatures. FMEA depicts the parallel and sequential combinations of events that cause a 'top' level fault to occur and then FTA can be used to determine what sensors are necessary to monitor the top-level event. The exact fault can further be visualized through fault tree diagnosis and also to highlight the critical signals to be monitored. For example, the cooling system of an automobile can be taken to develop a dynamic model and transfer the velocity command as input and the motor current as output. The series of experiments were conducted at different speed and data were collected, which were later on analysed to fit in a particular model. It was observed from the study that system monitoring is capable of monitoring changes in the observations made and therefore, reduction in number of sensors being used.

BENEFITS FROM CONDITION BASED MAINTENANCE

The following benefits can be expected with the use of condition based maintenance:

1. **Safety:** Introduction of safety measures in equipment/system can lower the contents of injury and reduction in fatal accidents. It enables system to either stop or to indicate the likelyhood of the expected failure in the form of alarm.
2. **Extended useful life:** With regular monitoring of equipment/system the unscheduled failures are minimized and therefore the useful life is extended.
3. **Enhanced availability:** Improvement in availability is possible if the unscheduled breakdowns are minimized. It is experienced that proper maintenance/up keep of equipment lowers the down time and therefore enhanced availability.
4. **Reduction in maintenance time:** As the unplanned failures are minimized with the use of condition based maintenance the repair time or maintenance time to correct the faults in components/parts is also reduced.
5. **Improved output:** The output of any equipment/system is directly related to its availability therefore with the application of condition based maintenance the output of it has to improve. Some of the machines are allowed to run at increased load because of its good health. The efficiency of such system also improves.
6. **Quality product:** With better health of equipment/system by the use of condition based maintenance its product quality will also be better. The services provided will also be of high standard to the satisfaction of users.

ADVANCES IN CONDITION BASED MAINTENANCE

Development of 'intelligent' electronic saving devices have brought in revolutionary changes in condition-based maintenance and automated production systems. The concept of predictive condition monitoring, to enhance the performance of machines and maintenance personnel, with the application of suitable information processing analysis and surveillance techniques is only a possibility today. On-line data acquisition and analysis has made it possible to predict the impending failures in machinery/system to a greater extent. The following are a few areas where advances have helped the condition based maintenance.

Sensing Devices

In today's world to manage an unmanned production system, the efficient gathering and processing of information is essential requirement of a control system and its associated software. Accurate real time knowledge of both system and the process is necessary to optimize the overall performance of machinery and therefore maximize the production. To run a machine/system uninterrupted it is important to monitor its condition/health regularly. Data acquisition is therefore a key requirement in the type of technology. Recent developments in integrated circuit technology have progressed to the point where the assimilation of various sensor functions in a single unit has been achieved. Information from several sources may be combined within the package to form one unit which may be named as intelligent sensor based on very large scale integration (VLSI) technology. The intelligent sensor may have some data evaluation capability to improve the signal conditions.

To develop a self-optimizing machining system Real Time Actual Condition Knowledge (RTACK) information is required in several areas with effective sensing systems. The areas of interest with respect to the development of intelligent sensor technology include tool wear and breakage, component size, surface finish alignment position and location, and the machining system. The tool failure must be sensed properly before its replacement. Similarly, true in-process measurement of dimensions and surface finish is difficult and therefore needs attention during unplanned machining.

Knowledge Requirement

In a unmanned machining system, human supervision has to be replaced with a sensor, which can acquire RTACK data, which can be applied in the following areas:

(*i*) Multidimensional sensors must be applied in place of individual sensor being used in control of a system.

(*ii*) Grouped sensors either in series or parallel configurations be used to control the information.

(*iii*) Multi-layered sensor input at various levels in a control system must be used to achieve optimum performance.

The key element in the concept is that the resulting information, essentially RTACK data, is used within an intelligent adoptive control system to optimize the machine operation in total and not simply to achieve sub-optimal goals. At present there are a few available sensor system, which are capable of providing RTACK data during machining operation. In some areas, non-contact optical sensors are being used to measure the surface finish. For in-process alignment positioning etc. vision or fast three dimensional laser scanning systems, appear to offer effective solution if the same can be developed.

Management Process

For the control of various operations in an automated system including maintenance, the management of all such activities is a challenging job and therefore, must be given due attention. Some software's are available but lot of work has to be done to overcome the control problems. However, as the computing and sensor power increases, the available techniques may be viable through development of intelligent machining systems. To form a smart machining system, an integrated approach that can incorporate, adoptive control with use of condition monitoring may yield the desired solution.

QUESTIONS

1. Highlight the importance and utility of condition based maintenance.
2. Discuss the role of condition monitoring in a computer based maintenance.
3. Explain various condition monitoring techniques used in practice.
4. How system approach to condition monitoring can be useful? Explain.
5. Illustrate advances being practiced in condition based maintenance.
6. Discuss the parameters on which the solution of condition based maintenance rests.

15

Computers in Maintenance

INTRODUCTION

Maintenance is a vital part of any organization and must be carried out properly to keep the equipment/system operational as far as possible. For effective discharge of maintenance function, a well-designed information system is essential. Such systems serve as effective decision-support tools in maintenance planning control and execution.

Maintenance planning is needed for preparation of maintenance schedules and job distribution. The need of manpower planning, at this stage can not be over looked. The planning must take care of organizational goals and policies and then formulate the procedures for effective implementation.

Today there is an increasing trend of application of information based decision supports in different departments of the modern organizations. Thus, computers have become indispensable requirement in maintenance management as well.

For optimum maintenance planning and scheduling, a large volume of data pertaining to men, material, money and equipment is normally required to be handled, which is difficult to be performed manually. For advanced maintenance systems adequate and timely information is must, which is only possible by the use of computers.

The computer is an efficient and reliable tool for maintenance personnel to plan and implement their programs. However, the success of a computer-based system solely depends on the way the system is planned and the way the data are processed. The inputs to the computer system are exactly the same as the inputs to a man-made system. The only difference is the storage capacity and processing speed. The work, which takes hours in manual operation, can be done in a fraction of seconds by the use of computers. Here, programs can be prepared to have the available inputs processed with the help of computers. Such a computer-based system can be used as and when required for the maintenance work to improve its performance.

COMPUTER-AIDED MAINTENANCE

In certain sophisticated and complex industrial organizations such as the power plants, fertilizer plants, petro-chemical plants, food processing plants, computer aided manufacturing, etc. temporary stoppages and shutdowns can be very serious in nature, and therefore, attention must be focused on systematic maintenance planning and scheduling. Though the planning is quite complex in nature and needs, an objective approach in decision-making execution and control. For these, it is desired to have accurate and up-to-date information in an organized manner. The industries mentioned above have a very large number of complex systems, thus, it is very difficult to monitor their maintenance requirements. Without a proper computer based maintenance system, the decisions made may be sub-optimal and erroneous. For this purpose, there are wide variety of software packages available in the market for different types of maintenance systems.

Computer enables implementation of on-line monitoring of important equipment/system. This on-line concept is widely used in condition monitoring, where the equipment performance is checked at a fixed interval and any deviation in performance is investigated for verification of fault. Thus, it helps in avoiding sudden and untimely breakdowns. In case of flexible manufacturing system or an automated system, some maintenance functions such as change of a broken/damaged tool, can be done automatically with the program provided in the computer system. With the technological developments in the field of artificial intelligence and expert systems, the risky maintenance jobs such as aero-industry or in steel plants can be attended with the help of robots.

A computerized maintenance system takes care of the following aspects:

- Development of database to support utilization of periodic maintenance program.
- Analysis of past records/failure history, to ensure the frequency of maintenance program.
- Development of maintenance schedules, procurement of maintenance material and facilities.
- Feed back control system to assess the feasibility of the programs to be used and monitored.
- Project management to achieve the set maintenance objectives in line with the organizational objectives.

Development of Database and Analysis

In maintenance system information is required about the failure of components/parts. This can be used to estimate the failure frequency. The data collected must be fresh and original. Their suitability, reliability, adequacy and accuracy must be ascertained. Because all other decisions will be based on the information provided. During data analysis, if the failure distribution is not clear, non-parametric tests must be carried out to test the hypothesis.

In case where information is not available, the same can be generated, following a particular distribution as applicable to the situation. However, rigorous tests are to be carried out to establish the failure/repair time distributions to match the real life situation. Otherwise the results obtained will not be authentic and accurate to be used in real life situation.

Development of Maintenance Schedule

With the available information, now it will be possible to generate maintenance schedules to know the intervals of maintenance functions. During this stage only, the planning for maintenance manpower must also be carried out. Since the nature of maintenance function is unpredictable due care should be taken to account for sudden and random breakdowns, which cannot be avoided with any type of maintenance policy. During schedule generation, the concept of backlogs jobs, which are not completed in due time, must also be kept in mind during manpower planning stage. As far as possible, the concept of dynamic scheduling must be applied to overcome all practical difficulties during planned maintenance programs.

Procurement of Maintenance Material and Facilities

The maintenance work generally suffers due to non-availability of spare parts and facilities in time. Forecasting spare parts and other relevant materials can be improved by the use of computers. The details of spare parts can be stored either by equipment-wise or item-wise. With the help of available forecasting techniques, the exact requirement of spare parts and facilities can be known in advance. If the same work is done manually, it will consume a lot of time and also the information may not be accurate as desired. The use of computers helps the management also to get the information quickly and correctly. The information from a central store to site stores can also be made available through Internet for speedy action. A material master file is created for all equipment/system under consideration, to keep the record of each item, and these records can be updated as per the changes made.

Spare Parts Life Monitoring System

Under this system, information about spare part such as description, anticipated life and the date of its installation in an equipment, and so forth, is usually recorded. As and when a particular spare part is replaced during breakdown or scheduled maintenance, the updating of this information is recorded in the respective files stored in the computer. With this information, it helps to prepare the following reports:

- Spares repeatability of same item in various machines, indicating the performance of such spare parts.
- Comparison of the actual life of the spare parts used, with the estimated life as recommended by the manufacturers.
- Maintenance of stock/records as per the requirements.

Spare Parts Tracking System

The total time required to rectify a breakdown of a machine, is the sum of total of the time, required for each of the following operations:

- To identify the failure with its cause.
- Time to determine the requirement of spare parts.
- Time required in procuring the spare part.
- Time required in rectifying the failure/fault.
- Time required in testing and certification of the equipment/system.

In most of the cases, maximum time is consumed in identifying fault and procurement of the spare parts. In a computerized maintenance system, the spare parts tracking system, is beneficial in this respect to a great extent. A spare part file is created that contains the information about the material code; spare part identification number, the assembly or sub-assembly number and the equipment/machine where this spare part is used. This helps in knowing the current position about a particular spare part and facilitates, timely procurement for future demands.

Material Requirement Planning (MRP)

By maintaining the information of spare parts in computer, makes it possible to place an order for the same, in time. The requirement of all maintenance related material could also be done effectively. The past information about the consumption of spare parts makes it easy to plan the future requirement with maintenance related jobs. The other facilities required can also be planned in advance since their time period is known. This also helps in availing the services of the experts in the area of maintenance for complex and costly equipment, where downtime of the system is expensive and may affect the goodwill.

Feed Back Control System

It is essential to prepare a job card for each equipment/component to record the maintenance work carried out or to be carried out. The job card shows the plant code, equipment code, the job code, the nature of job, the time of initialization of the card, the start time of the job, the finish time of the job, the man-hours spent and so forth. These filled cards are sent to computer section for processing and necessary compilation. The feed back control system helps the maintenance engineer for compiling the following information:

- History of maintenance programs carried out.
- Frequency of maintenance operations.
- Average maintenance time for various jobs taken to complete them.
- Backlog of job, rescheduling requirements.
- Cost involved in maintenance functions equipment wise.
- Manpower utilization category-wise.

The above information can be used to carry out modifications, so desired and to plan the maintenance activities in such a way to make proper use of facilities and material. It will also help in:

- Providing accurate quantity of spare parts.
- Reduction in non-moving lot of spare parts.
- Maintaining the optimum level of spare parts inventory.

The other important responsibilities of the store management are the following:

- Receipt orders
- Processing of orders
- Procurement of spare parts
- Inspection of store items for quality assurance
- Issue of store items to sub-stores
- Stock record keeping
- Valuation of stock

Project Management

The all above stated problems/functions are to be managed by a group for efficient working. Sometimes conflicting situations arise, for example, the cost of maintenance vs availability of services for human cause. This must be resolved amicably without burdening either the user or the organization concerned. For the purpose of management, the use of computers can help to greater extent, where prompt decisions can be taken based on information available.

RECORDING OF FAILURE STATISTICS

For effective and efficient maintenance functions relevant information is essential and can be obtained through various agencies. Before using this information, it desired that the same may be scrutinized and be presented in the proper format to highlight the significant features of the data at a glance. From failure data, the frequency distributions must be constructed to know their nature, which can be applied for analysis. Where, data are not readily available the same can be generated as per the specified distribution.

Statistical analysis has to be carried out to validate the data to yield realistic results. This all can be easily done with the help of computer since a large amount of data is to be handled and processed. Besides the processing of data in computer, it is easy to store them also. The confidentiality of the information can be maintained, as well as transfer is also possible through Internet without any paper work. In the cases where, condition monitoring is practiced with the dedicated system collection and recording of information becomes relatively easy.

With the availability of modern techniques feeding of information to the computer is made simple and time saving. It can also be corrected/modified without much difficulty. The data can be formatted in the desired form for presentation or for report preparation. The statistics so developed will be used to develop maintenance program and schedules. The requirements of spare parts can also be computed to help their procurement. While computing the spare parts, the following points must be considered:

- The cost of downtime of equipment/service.
- The cost of stocking the spare parts.

But in practice it is difficult to maintain proper balance of the above two factors and the following steps must be kept in mind:

- Efficient indenting of spare parts.
- Control over consumption.
- Reconditioning/overhauling of used spares.
- Use of management techniques to control maintenance related activities.
- Effective involvement of maintenance and operating personnel.
- Obsolescence of equipment/machines.

COMPUTERIZED MAINTENANCE MODELS

In the systems where condition-based maintenance is to be followed, the computers play a significant role. The role of real time accurate knowledge is very vital for maintenance function, as well as to optimize the overall performance of the equipment/system. With the help of data acquisition system (DAS) the condition of an equipment/system can be monitored at a fixed interval of time and accordingly the maintenance schedules can be drawn and implemented. Other tools such as shock pulse meter, vibration analysis can also be used to assess the condition of machines. In a continuous process/system, on-line monitoring is carried out to monitor the health of the machinery. In these systems, warning switches are also provided to indicate the impending failures. This is used from safety aspect also.

When the number of critical parameters is more or when the rate of monitoring the critical parameters is high, the use of computers is most desirable. The critical parameters such as vibration, temperature, pressure and so on so forth, which need to be continuously monitored, can be recorded on the on-line monitoring system, interfaced with the computer.

Through this data, the base line signature of the particular component/part can be maintained in the computer file. The signals received from working systems can be compared with the base line signature. Any deviation in the measured value of the parameter beyond the tolerance limit is highlighted for necessary remedial action. This total procedure is called signature analysis.

Computer based maintenance systems can be implemented for:

- Monitoring critical parameter through data acquisition system
- Monitoring critical parameters through feed back system
- Establishing breakdown analysis system

When several similar equipment/machines are used in any industry, the computerized breakdown analysis helps in processing the inventory and planning the preventive maintenance schedule. This system aims at generating important details for the purpose of breakdown analysis and updates the equipment history as well.

Equipment log sheets are used for collecting information about the equipment performance, breakdown durations, and reasons for breakdown and corrective actions taken. This information is updated at regular intervals of time. A master file is maintained to know the spare parts requirement.

Information is power and should be utilized properly for any system and particularly for the maintenance system, since the success and failure depends upon the accuracy and timely availability of the information. This information helps in taking decision if collected properly and presented in the desired format.

DECISION-MAKING IN MAINTENANCE

Maintenance management involves critical decision-making at various levels. Decisions related to operations and overall economic performance is vital for accomplishing the objectives of maintenance. The maintenance department of any industry works in association with other departments. Interaction with other department must be well defined to avoid conflicts and achieve smooth functioning of the system. Some of the important decision-making process in the maintenance department is concerned with scheduling of routine servicing and inspections, minor repairs, temporary shutdown of a system for major overhaul and repair and replacement of aging and obsolete equipment. Thus, it is essential for the maintenance manager to understand the basic method of decision-making.

The decision required to be made by the maintenance department concerns about the following points:

1. Identifying the maintenance objectives and priorities of the jobs, depending upon their requirements.
2. Selecting the most appropriate and suitable working method out of the available alternatives.
3. Maintaining the reliability and failure rate data of components/sub-assemblies/systems etc. at minimum level.
4. Predicting the impact of external factors that are beyond the control of the decision maker.

The various steps required in decision-making are the following:

- Identification of the need for decision-making
- Identification of problem and influencing factors
- Identification of the alternative course of actions
- Action to define methodology
- Classification of the influencing factors

- Define the objectives
- Define the decision criteria

Decision-making is a complex problem and some guidelines must be kept in mind while framing information systems for the maintenance function. The following points need attention:

- Information collected must be flexible to allow evaluation of the many alternatives.
- All alternatives must be able to accommodate changes because of diverse human nature and values.
- They should be capable of being applied for analytical work and serve as intuitive models for evaluation, keeping track of many alternatives and consequences.

Efficiency of decision-making depends on the quality of information based on which a decision is derived. Thus for effective maintenance management, it is essential to design a system such that all the relevant information flows to the maintenance department. Such information must be processed with the right perspective prior to any decision-making. The maintenance department utilizes information for the following purposes:

- Building up intelligence in maintenance related activities.
- Information for design phase after the system has been used and its demerits are identified.
- Information for selection of alternatives.
- Information for implementation phase.
- Information for maintenance phase.

Decision Models

Organizations are made up of individuals and as individual thinking and priorities vary from person to person, conflicts may arise in taking a decision. Therefore, it is desirable to build information systems to facilitate the decision-making process effectively. There are various models of decision-making. A few of them are discussed below:

Rational Model

Some economists, mathematicians and management experts feel that decisions are made rationally because of the following reasons:

- The decision makers know the objectives and their importance in the order of their merits.
- They also bear in their mind that the alternatives to each problem, with their merits and demerits.
- They select the alternative, which is likely to yield maximum profits and meets other objectives.
- They implement the selected alternative to yield the desired results to satisfy the objectives intended.

However, in case of maintenance function, the alternatives are limited and the decision has to be taken purely on the importance and requirements of the equipment/system. Decision-maker does not know sometimes, all the alternatives and their consequences. In spite of the limitations, rational models are powerful tools for decision-making.

Satisfying Approach Model

Simon has proposed a model based on satisfying approach rather than optimizing approach. In this model, the satisfaction factor is given much weightage. To know how much satisfactory will be a particular alternative, extensive information about all the possible consequences of all the alternatives would need to be enlisted.

It may not really be possible for the decision maker to examine all the alternatives with their consequences as suggested in this model whenever possible, the new uncertain alternatives are avoided and testing rules are adopted. Some of the alternatives may be conflicting in nature. Therefore, an alternative, which satisfies the functional and maintenance requirement should be selected.

Muddling Through Model

Lindblom (1959) suggested a radical deviation from the rational model of 'successive limited comparison'. In this case, values are chosen at the time when policies are framed. For instance, labour and management may not agree on values but they can agree on policies. Because of the limits of the human rationality, Lindblom suggested incremental decision-making or choosing policies keeping in mind the comparison with the previous policies. Decision-making is continuous process in which final decisions are always modified to accommodate changing objectives. In the case of maintenance, the objectives do not change very much. Therefore, any decision is treated to be final at the very outset.

Limitations of Failure Statistics

There are various methods of failure analysis. However, a mere knowledge of these methods does not help if reliable data is not available. Acquisition of such authentic data for both human and machine factor is always difficult task due to resistance from the working personnel. There are a few methods for the collection of these data, but care must be exercised. The data acquisition system should therefore be designed in such a way that it remains simple and does not miss out any vital piece of information. A simplex format must be developed to provide correct information at the appropriate time. The information can be collected at different levels.

Departmental Level

Once the need for the information is recognized, it is easy to list out items of information to be used as database at the departmental level. The first information required is that of identifying the failure prone components together with their frequencies. Once the nature of failure is known, it is easy to plan for manpower and material accordingly. This information can be collected from the history card of the equipment/system.

Organization Level

Having collected information at departmental level, the same can be used to decide the replacement policy or to optimize the maintenance cost of the whole organization. The centralized information at the organizational level can also be helpful to the manufacturers to improve or modify their products.

Public Enterprise Level

There can be public enterprises providing services of information processing for maintenance planning and implementation. The main aim of such enterprises is to collect information from the organization and compile it, based on the type of equipment and its utility. These agencies receive information in the form of reports forwarded by the client organization. The reports contain present maintenance status and degree of its compliance as per the targeted goals.

The task of the information-receiving agency is to give the feed back by comparing the actual maintenance carried out with the scheduled maintenance work. This practice helps in making an estimate of the maintenance cost for comparison of the same with the production cost. The other utilization of this information is to help the upcoming industries to assess their status vis-à-vis established organizations.

Use of Pooled Information

The information available at central place can be useful at both the departmental level as well as at the organization level. For example, if a centralized inventory is maintained, the department or the organization can know the availability or non-availability of a particular spare part.

When an up-to-date information system has been developed for the maintenance function, a new area of research in maintenance can be initiated, which will definitely help the organization. Indigenous materials/processes have to be evolved to sustain the growing demand for new equipment/components.

COMPUTERIZED MAINTENANCE PLANNING

Computers can be applied in the area of maintenance planning as they are used in other areas. Maintenance planning is used to optimize the use of resources available. The manual work of recording various aspects of maintenance such as work order, priority of work, material planning, work plan, job sequence etc. can be handled carefully and accurately with the use of computers. It is therefore a powerful tool for assisting maintenance planning. A well-conceived computerized maintenance planning system should provide manpower back logs equipment histories, equipment parts list, material availability, preventive maintenance schedules and associated costs.

Effective maintenance planning requires observing planned job progress to eliminate potential delays or problems that may arise. During the following up process the planner may address problem areas which are:

- Were the communications clear and adequate for the personnel involved in the maintenance work?
- Was the job plan simple enough to understand?

The basic reason for planning maintenance is to eliminate the cause while overcoming the effects efficiently. The benefits of computerized maintenance planning include greater access to information, better planning of resources, increased control and finally reduction in overall cost though the initial cost of the system may be higher which can be compensated in due course of time.

QUESTIONS

1. Discuss the role of computers in context with maintenance functions.
2. What is the computer aided maintenance? Explain.
3. Explain spare parts life monitoring system in brief.
4. Highlight the importance of recording failure statistics for adequate maintenance.
5. Discuss various computerized maintenance models used in practice.
6. Highlight the importance of decision making in maintenance.
7. Explain the limitations for recording failure statistics.
8. How computers can be useful in maintenance planning? Explain it briefly.

16

Organizational Structure for Maintenance

INTRODUCTION

Technological growth worldwide has brought out significant changes in equipment and machineries. The new concept of complexity in design and manufacturing is possible by the use of computers, artificial intelligence, knowledge based expert systems. These phenomenal changes have necessitated the optimum utilization of these sophisticated equipment/systems. To keep the equipment/system operational, the demand for maintenance has also increased manifold. With this in mind, development of organizational set up for maintenance is of paramount importance. Before setting up a new maintenance organization, its need, objectives, functions, etc. must be understood clearly to make it a cost effective proposition. Nowadays, the concept of total productive maintenance (TPM) is gaining momentum, where, workers are not divided such as maintenance and production personnel and each one of them is capable of undertaking any job. This adds flexibility in the system as the normal conflict of production and maintenance personnel is avoided.

OBJECTIVES OF MAINTENANCE ORGANIZATION

Maintenance of modern machinery and industry calls for setting up of a healthy, balanced and rationalized organization to achieve the multi-faceted objectives. This is essential to keep the diverse activities of the maintenance department under control. An effective maintenance organization can be set up after careful study of the background of the industry. While setting up the organization, the future trend of the industry should also be assessed so that any modifications in future in the set-up, can be incorporated.

The importance of the maintenance organization grows with the increasing size of the industry. A small-scale industry may require a simple maintenance organization with a foreman in-charge, whereas, if the size of the industry increases, the complexities of the maintenance and, therefore, the maintenance operations, functions and organization also grow in size. A highly trained and competent person may be required to head the maintenance organization/department. There has to be a close relationship between the production and maintenance departments.

The basic objective of any maintenance department/organization should be to ensure that the maintenance function is carried out effectively and the equipment downtime as well as the loss of production is minimized. High equipment availability can be achieved by proper coordination of maintenance functions with other departments.

The head of the maintenance department/organization should possess the necessary expertise and experience since the success or failure of maintenance department would depend upon him. He has to look after not only the day-to-day maintenance/repair of equipment but also ensure the lifeline of the production unit. High availability of production system can be obtained by continuously monitoring the planned preventive maintenance schedules identified for the machine/equipment. In a highly mechanized and sophisticated organization, the chief engineer (Maintenance) should normally head the maintenance department. The structure of the maintenance department/organization depends on the type of the industry as well. For a complex industry, for example, mining, a centralized maintenance unit can be set up since hundred of mining units operate independently in different areas. In such type of organizations a maintenance unit can be set up at the headquarters, and several field branches of such a unit can be set up at convenient places near the working areas.

The following points should be considered while setting up a maintenance organization:

- The organizational goals pertaining to the maintenance function must be identified.
- Maintenance workload should be determined before hand to achieve the desired results.
- Total maintenance work should be properly and uniformly distributed among all concerned persons to avoid conflict among the workers.
- The essential work should be assigned clearly and definitely to various departments of the maintenance organization.
- Due consideration should be given to proper qualification of the workers while assigning a particular job.
- The maintenance organization should be staffed with the best qualified and trained people available.
- The policies and procedures should be clearly designed at an early stage to help maintenance people to achieve the organization goals.

In developing maintenance organization, it is attempted to compromise between the individual needs and the organizational goals.

MAINTENANCE FUNCTIONS AND ACTIVITIES

A good maintenance organization should reflect the total maintenance activities required to be undertaken by the various sections of the department. The functions and activities associated with the maintenance organization are as follows:

- Identifying areas/systems for implementation of preventive maintenance program.
- Making suitable arrangements for maintenance facilities for carrying out the maintenance work properly.
- Planning and scheduling of the total maintenance work of the organization including breakdown maintenance.
- Ensuring proper and timely supply of spare parts through an efficiently managed inventory control system.
- Carrying out standardization of maintenance work to minimize down time of equipment/services.

- Arranging import substitution of components, if necessary.
- Implementing modifications to the existing equipment, wherever possible.
- Assisting the purchase department in procuring new machinery, construction work related to machines, and replacement of the parts of the equipment.
- Disbursement of services such as, water, compressed air, steam, electricity, and the like.
- Identification of obsolete and surplus equipment for replacement and disposal.
- Training of maintenance personnel.
- Ensuring safety of personnel and equipment.

ORGANIZATIONAL REQUIREMENTS

While setting up a new maintenance organization, the following parameters must be kept in mind:

Coordination

Success of an organization depends on effective coordination and cooperation of the people in the organization. For a prolific maintenance organization it is essential to have a clearly defined organizational chart indicating job description and unambiguous definition of authority for each job. A group, rather than an individual performs most of the maintenance jobs. Therefore, it is utmost necessary to maintain work harmony. To maintain proper team spirit, the leadership of the group is very important which can bring people together. Thus, success of maintenance function entirely depends upon the dynamic leadership of the departmental head.

Maintenance Engineer

The role of the maintenance engineer is very important and crucial, as his primary job is to reduce the downtime of the equipment/system under his supervision. Therefore, the maintenance engineer should be a well qualified, technically competent, knowledgeable and cost conscious in order to be an effective and efficient leader of the maintenance group. The output of the maintenance shop, vis-à-vis the time, talent and effort put in by the team, is the index of its efficiency. Application of proper management techniques, motivation of workforce, avoidance of conflicts and inspiring leadership of the maintenance engineer can bring the targets and achievements closer.

Attitude of Personnel

The attitude of the working personnel in many instances is not always positive towards accepting the modern facilities/techniques available for the maintenance function. Under the existing conditions the maintenance personnel should be apprised of the latest techniques available in the field of maintenance together with their merits and demerits so as to change their mindset and encourage them to adjust to the changing environment. This will also help the organization to achieve its goals. Organisations must correspondingly frame their policy to bring change in the attitude of their personnel towards acceptance and implementation of advanced maintenance concepts.

Policy

The policies adopted by the maintenance organization should be such that they can be easily understood and followed by everyone within its framework. These can be either formal or informal. The formal policies are recorded and are subjected to modification from time to time. It is the responsibility of the maintenance boss to frame the policies, in consonance with the company rules, considering working hours, benefits, safety, security, overtime, and the like. The policies framed should fit in the organization set up as well. However, informal policies are practiced, as and when required to avoid clash amongst the workers.

Maintenance Work Control

The structure of a maintenance organization should be such that there is always sufficient manpower to supervise the work. Care should be taken to see that a balance is always maintained between the supervisor and the number of workers to be supervised. However, there is no fixed rule to maintain the proper ratio of supervisor to workers, although effective monitoring is most essential. The supervisors must understand the problems of the maintenance group, due to criticality of certain maintenance work, which may be time consuming; sometimes detection of fault itself consumes considerable time, which may annoy the supervisors.

Managerial Functions

It is the duty of the supervisory staff to develop their subordinates to a required level of proficiency. The working personnel should be properly motivated to attain high levels of proficiency so that they can work freely, fearlessly to fulfill the organizational objectives. The development of the subordinates can be done keeping the following in mind:

- Level of education
- Level of skill
- Level of intelligence
- Level of motivation

It must be pointed out here that the above levels entirely differ from one person to another. The jobs to be handled also require different levels of skill, such as unskilled, semi-skilled and highly skilled. This should also be borne in mind during the development stage of the subordinates. In this regard, the leadership of the supervisor plays a very important role. To be a good supervisor one should be a man with appropriate information and knowledge of the subject and also a very good human being to motivate the workers.

Capacity Utilization

The main function of the maintenance organization should be to utilize the full potential of its workers. During the selection of persons, the values of each individual must be assessed thoroughly, as it is seen that a person may be a very good engineer but a poor manager. However, managerial qualities can be imparted with proper training and incentive and thus a good engineer can become an efficient manager as well. It is also sometimes noticed that individuals move from one organization to another for better career opportunities. This can be taken care of by the top management so that the skilled and efficient persons are retained in the organization. The real leadership calls for an effective utilization of the capacity of the individuals. This can be best achieved by properly assigning the jobs to the personnel. The maintenance work should be projected as a challenging job so that persons, who are competent and dynamic, do not hesitate to take up this assignment. The persons, who are willing to undertake the maintenance work, should be considered for the assignment to keep up their interest.

ORGANIZATIONAL PROBLEMS

Maintenance organizations face various problems due to the nature of the maintenance work. To overcome these problems, the maintenance organizations, with the cooperation of the plant management and the production department, should try to fill up the organizational gaps arising out of the following situations:

(*a*) The plant management must identify the objectives of the maintenance organization with overall production strategy. Sometimes the production department does not allow the maintenance personnel to take up the scheduled maintenance work as this disturbs their production targets.

(*b*) The maintenance work is of varied nature; the maintenance organization should understand the difference between the routine and essential work so that the essential work is carried out most expeditiously with the help of the production department.

(*c*) A clear distinction between field and shop work should be made so that this traditional attitude does not hamper the maintenance work. The maintenance organization should try new concepts for better work control procedures.

As stated earlier, properly qualified and experienced persons should only man the maintenance organization. Sometimes persons from other departments are also shifted to the maintenance department either on promotion or due to their unsuitability in their respective departments, which should be avoided as it may create serious problems in the maintenance organization.

It should be borne in mind that, though the policies set up by the top management should act as guidelines for every activity of maintenance function, the actual procedures are the outgrowth of such policies and must be given flexibility to accommodate changes, if anticipated.

It is normally seen that if more than the required number of people are engaged in a work, it ultimately suffers, as nobody owes its responsibility. Sometimes, too many assistants in a particular job create confusion among the workers due to multiple chains of command and this must be avoided as for as possible.

During the planning stage of an organizational set up, it may also be remembered that over-staffing needs to be avoided to obviate overlapping of duties, at the same time sufficient manpower should be provided to complete the maintenance related jobs.

TYPES OF MAINTENANCE ORGANIZATIONS

The selection of a particular type of system will largely depend upon the main structure of an industry. Maintenance organizations, however, can be of the following types:

- Decentralized
- Centralized
- Partially decentralized

In large-size plants located at different places, inter-unit communication is difficult. In such cases the decentralized type of organization is best suited, which means that the maintenance organization works under the direct control of a chief engineer in-charge of production. Under this type of organization, better coordination between the production and maintenance group is possible because of a common head of organisation. The advantages of such type of organization are:

- Speedy decision due to better line of communication under single control.
- Maintenance and production people understand each other's problems better because of their common goals.
- Interchangeability of work force, even at managerial level, is also possible.
- Better training at the workers level can be arranged.
- Conflicts among workers are also reduced due to better coordination and cooperation.

In a small factory where communication between the departments is more free, the centralized type of maintenance organization is preferred, which is placed under a chief maintenance engineer/manager. The total responsibility of maintenance function, for the entire factory lies with the chief maintenance engineer. Under this type of organization, the responsibilities and accountability of work must be properly specified for production as well as for maintenance personnel to successfully meet the project goals. If this is not taken care of, one department may blame the other for any shortfall.

The partially decentralized organization, which is the modified form of a centralized type of organization, is most suitable for projects that have units at far away locations. Under this type of maintenance organization, the maintenance personnel, attached to the production unit carry out work at unit level for day-to-day maintenance. However, important maintenance functions such as overhauling, planned maintenance work, procurement of spare parts, etc. are kept under the charge of a chief maintenance engineer at his central office. All centralized work as well as maintenance planning, and documentation are done at the central maintenance workshop.

The above three types of maintenance organizations however, are not strictly followed and some adjustments can be made to suit the working environment and the need.

Basically there are two types of organizations that are followed in most of the projects/ industries – line organization and staff organization. In the past, the line organization was mostly followed, which consisted of a general foreman and a number of specialist foreman with their own working personnel under them. Here, the specialist foreman executed maintenance work in their respective areas while the general foreman supervised the total work under his control. Figure 16.1. shows a typical line organization.

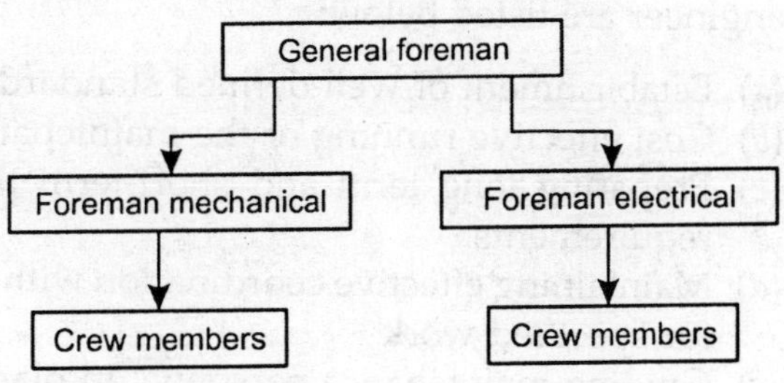

Fig. 16.1: Line Organization

Realising the need of recording the maintenance work, a few staff members when introduced in the line organization, it forms a line-staff organization (Fig. 16.2).

In the recent past, the maintenance organization was based on the craft concept. The persons used to join the organization as apprentices and rose to higher positions (foreman) after accumulating sufficient experience in their jobs. However, these persons were never given a formal training of the foreman job. Such an organization emerged emphasized craft skills and therefore strongly supported the idea of specialization gained through skills. Further growth in the field of maintenance function led to the concept of functional organization that was purely based on craft organization. In this system, a few workers are placed under each functional foreman (Fig. 16.3).

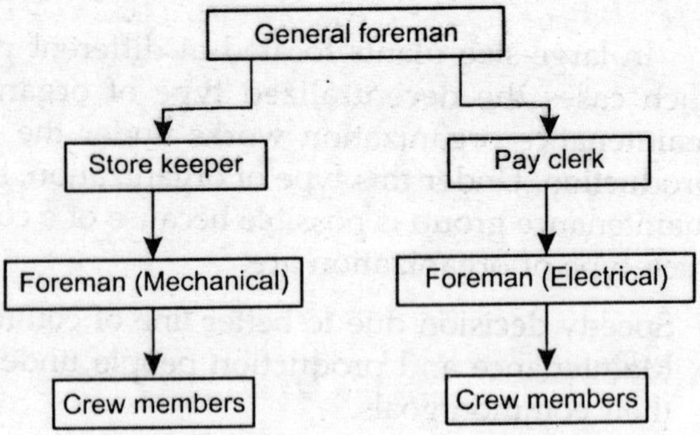

Fig. 16.2: Line-staff Organization

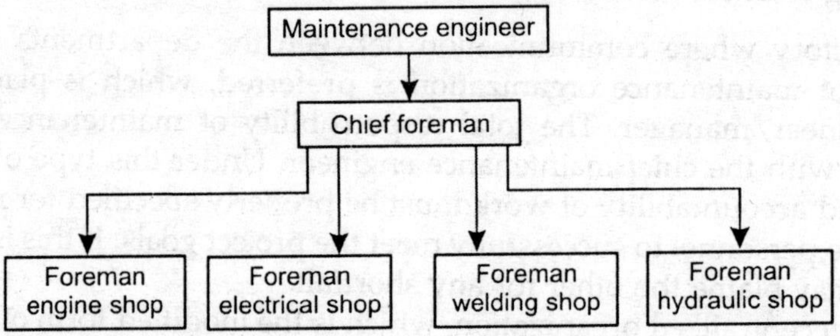

Fig. 16.3: Maintenance Functional Organization

In the mining industry area type of maintenance organization provides better utility of available manpower. Here a single person takes care of maintenance for a particular given area in a well-defined manner. He is provided with a number of workers to carry out day-to-day work including emergency work and to coordinate with centrally controlled shops for other important maintenance work. Because of less amount of maintenance work at area level, the utility of manpower is usually low but with better interaction with the central workshop, this problem can be eliminated to some extent. The duties and functions of an area maintenance engineer are listed below:

(*a*) Establishment of well-defined standards and practices for the maintenance work.
(*b*) Cost effective running of the maintenance organization.
(*c*) Preparing long term and short-term plans for area maintenance including manpower requirements.
(*d*) Maintaining effective coordination with the centrally controlled maintenance organization for specified work.
(*e*) Guiding maintenance personnel in planned maintenance work including its control.
(*f*) Maintaining proper documents of the maintenance work undertaken by his team, for future planning.
(*g*) Provisioning of spare parts for scheduled as well as breakdown maintenance.

The area maintenance organization, as shown in Fig. 16. 4 may be adopted for engineering services such as mining.

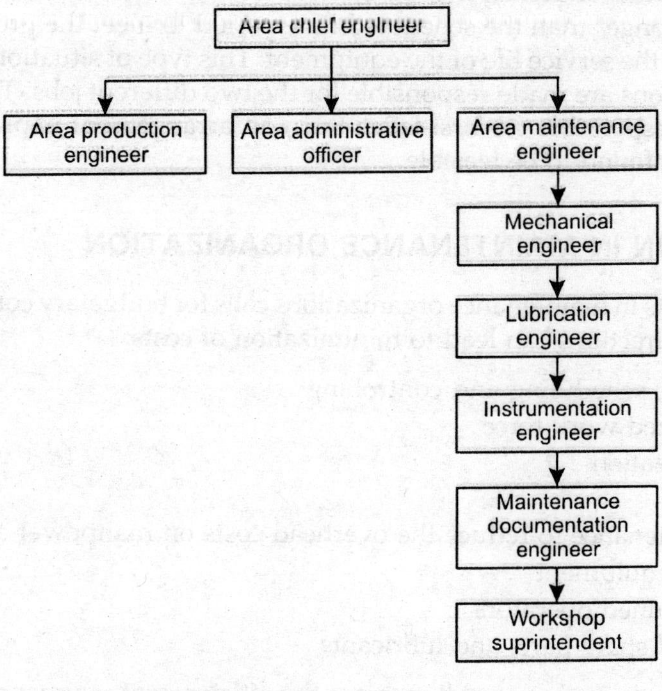

Fig. 16.4: Area Maintenance Organization

In some of the industrial organizations, the centrally controlled types of maintenance systems are adopted where all major types of maintenance work are undertaken by the central workshop. This is generally possible for all types of equipment, which can be shifted to the workshop. Moreover, such type of maintenance work can be taken up during the shutdown period of the plant. The maintenance superintendent, as indicated in Fig. 16.5, controls this type of organization.

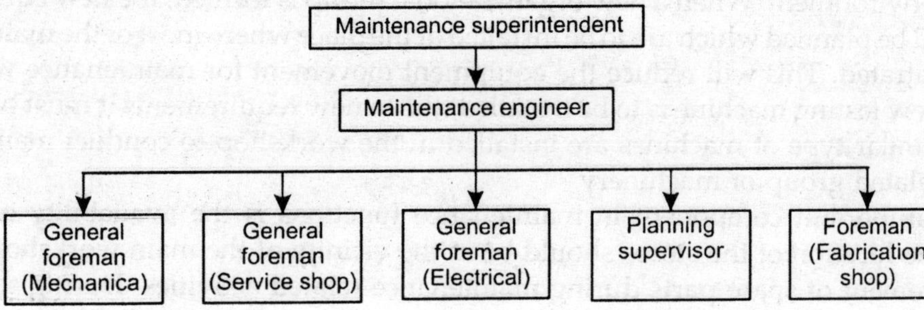

Fig. 16.5: Centrally Controlled Maintenance Organization

Some subjective type of analysis can be done to find out the merits and demerits of different types of maintenance organizations, for the final selection of one for a particular project/industry. For example, the production superintendent may also be in-charge of exercising control over the maintenance organization. But in big projects, a separate person may be doing this work independently. It is sometimes seen that the production superintendent utilizes the equipment longer than the scheduled time period to meet the production targets, which ultimately affects the service life of the equipment. This type of situation can be avoided if two independent persons are made responsible for the two different jobs. The operators are also sometimes made responsible for first-level. Such an arrangement is practiced in many industries and has been found to be feasible.

COST MINIMIZATION IN MAINTENANCE ORGANIZATION

To contain overhead costs in maintenance organizations calls for budgetary controls. Adequate steps in the following directions can lead to minimization of costs:

- Centralized planning, scheduling and controlling
- Grouping of specialized work force
- Effective labour utilization
- Budgetary control
- Use of contract maintenance to reduce the overhead costs on manpower and equipment
- Purchase of reliable equipment
- Use of skilled and trained operators
- Use of correct type of spare parts and lubricants

Due attention to the above points will increase the efficiency of maintenance functions at all levels.

UPDATING MAINTENANCE ORGANIZATION

Due to rapid advancements in technology, changes in the maintenance set-up have become essential. There should be flexibility in the maintenance organization to accommodate such changes. Special equipment are always required to speed up the maintenance work and therefore, the maintenance engineer must keep in mind their utility in view of the changing technological environment. When a new organizational set-up is formed, the new equipment/ facilities should be planned which are to be installed at the place where most of the maintenance work is concentrated. This will reduce the equipment movement for maintenance work. For example, if a new testing machine is to be installed due to new requirements it must be placed, where other similar type of machines are installed in the workshop to conduct maintenance work on the related group of machinery.

The most important component in maintenance functions is the availability of quality spare parts. The location of the stores should be in the vicinity of the main workshop to save time in procurement of spare parts during maintenance related activities. During scheduled major overhauls/repairs some extra facilities such as cranes, trolleys, special tools, etc. must be kept in place, so that, the same are easily accessible to the maintenance workers. In a big industrial organization all facilities should be made available to save time and money. Due attention should be given for the availability of qualified, trained manpower for maintenance organizations.

QUESTIONS

1. Discuss the role of maintenance organization on the performance of maintenance functions.
2. Discuss the objectives of maintenance organization.
3. What is the maintenance functions and activities associated with it?
4. Highlight the organizational requirements of maintenance functions.
5. Explain the problems faced during the establishment of maintenance organization.
6. Discuss the various types of maintenance organizations used in practice.
7. Point out the merits and demerits of different maintenance organizations.
8. Discuss the needs of updating any existing maintenance organization.
9. Elaborate measures taken to minimise the cost in maintenance organization.

17

Economic Aspects of Maintenance

INTRODUCTION

Optimal performance of machinery is a must for economic viability of any capital-intensive industry. Maintenance functions play a vital role towards achieving higher production targets, however, the cost of production operations must be within the laid down limits. To this end, every industry establishes a maintenance department to achieve their requirements effectively. One of the important areas of decision-making in the maintenance department is the economic consideration. Economic aspects of maintenance are looked into from different perspectives. Under certain conditions the cost of maintenance cannot be compromised seeing the utility of equipment and safety aspects. Here maintenance works has to be carried out irrespective of expenditures made on repair/maintenance of the systems. Aero industry and underground mining are the examples, where safety of men cannot be compromised at any cost. The prevailing methodologies include the life cycle costing, value functions, performance auditing, and the like. The economic analysis is also necessary for management of inventory and adoption of replacement policy. Sometimes it is observed that the maintenance cost rises very high as compared to the equipment cost but under the circumstances the availability of new facility plays an important role to take decision.

The success and effectiveness of maintenance functions depends on the care taken to estimate its cost accurately and in time. The labour costs can be controlled by cutting over time, regulating crew size and provide full maintenance load to all maintenance crew. Sometimes the non-availability of spare parts and facilities adds to cost component. Cost control must provide for the accurate allocation of costs to departments using maintenance facilities. Feedbacks from accounts section and details of conformity with departure from estimates and budgets must be taken for actual analysis.

LIFE CYCLE COSTING

It is the normal practice in industrial sector to purchase equipment based on the lowest quoted price. However, some organizations may purchase higher priced items, on the grounds of superior performance and low maintenance cost during operation. This may be an indication of the drawbacks of the said equipment. The low acquisition cost of the equipment may be one of the reasons for its higher operating and maintenance costs. Therefore, it is important that the selection of equipment be based on the total ownership cost over its entire life span. This total cost of ownership is calculated considering the operations cost as well as the maintenance costs over the product's life span. Such type of costing is called life cycle costing.

The concept of life cycle costing brings together engineering, economic and financial disciplines in connection with the equipment to be procured. The economic life of equipment depends on its maintenance and repair costs, and the resulting availability and operational efficiency. It can be estimated by plotting the cumulative efficiency and maintenance and repair cost per cumulative hour against operating hours. The interaction of two curves will indicate the economic life (Fig. 17.1) of the equipment/system.

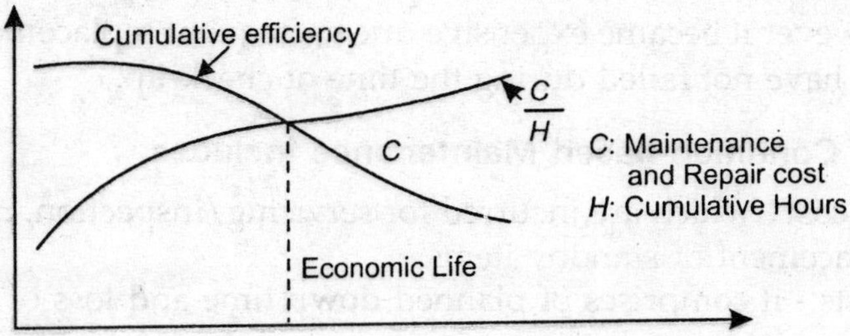

Fig. 17.1: Estimation of Economic Life of Equipment

The aim of life cycle costing is to know the total cost of equipment accruing over its whole life period, which may include all the costs starting from the specification cost. It is also observed that reduction in one cost may increase the other cost. The total cost has to be therefore optimized by making trade-offs among all elements of it. Normally, it is observed that cheap items do not have sufficient economic life nor do they yield satisfactory performance over their whole life, because of poor quality of material or process.

The various elements of costs involved in ownership of equipment are:

- Acquisition cost (at the time of purchase of equipment/system inclusive of all allied costs).
- Installation cost including commissioning and testing costs.
- Spare parts cost including the cost of all support materials such as test equipment, jigs and fixtures, and the like.
- Operating cost of the equipment including the cost of facilities, utilities operating personnel and their training.
- Maintenance cost including the cost of training of maintenance personnel.

The main advantages of the life cycle costing are:

- It may result in selection, of equipment, that has lower operating and maintenance costs and finally resulting in reduced cost of ownership.
- The money saved can be used for some other works or better quality equipment may be procured at the same cost.

Maintenance Cost

In practice it is difficult to control the maintenance costs due to unforeseen reasons of uncertainty in machine/equipment failure. However, such failure can occur in the form of premature or accidental breakdowns during operation time. The maintenance costs can be classified under the following categories:

1. Total of Breakdowns/Repairs

(*i*) This cost consists off finding out the fault/defects, replacements/repair and repair of standby components/their replacement.
(*ii*) Indirect cost in the form of production loss.

It comprises of unplanned down time, damage of equipment/system and quality loss/scrap. To overcome the above difficulties the concept of preventive/planned maintenance has come in to practice. However it became expensive due to frequent replacement of the components/parts which must have not failed during the time of check-up.

2. Total Cost of Condition-based Maintenance Includes

(*i*) Permanent cost - which are incurred for servicing/inspection, replacement/repair and repair/replacement of standby items.
(*ii*) Indirect costs - it comprises of planned down time and loss of production during this period.

3. Safety Related Costs

In some cases this is significant where safety of personnel/equipment is required by legislation or the requirement of the situation.

4. Special Maintenance Cost

This includes the cost associated for the improvement of plant, modification, modernization etc. for quality products.

During the planning stage, budgets are normally allocated for each type of activity, which must also include the maintenance cost. In the case of production the cost can be set easily. Whereas, in the case of maintenance function, it is difficult, since failures are random in nature and their levels also cannot be defined and therefore calculation of maintenance cost is very difficult in advance. The usual practice is to set targets for maintenance cost based on past experience. The maintenance cost, however, will directly depend on the level of maintenance and its requirements. It has been estimated that about 70 per cent of the cost of maintenance, spare parts and materials is incurred on account of replacement of components and assemblies subjected to wear and tear, and 25 per cent of this cost can be saved by systematic application of advanced tribology practices. Such improvements enhance the useful life of equipment by containing the rate of wear. Consequently, there is reduction in down time of the equipment; it improves the production as well. As a result, the demand for spare parts reduces and it becomes possible to manage the maintenance work with less manpower. All these factors have a great potential towards contributing overall reduction in the costs of the products. For maintenance work, the proverb 'a stitch in time saves nine' holds good.

The study conducted in a steel plant showed that improving the maintenance engineering practices might increase the profitability. This is possible through advanced maintenance practices. Table below indicates the gains realized through better maintenance practices.

Table 17.1: Ways of Cost Reduction in Steel Plants through better Maintenance Practices

	Major heads of maintenance cost	*Current cost as % of production cost*	*Cost reduction by 25% improvement of direct material cost*
1.	Spare and material for regular maintenance	4.75	3.56
2.	Lubricants	0.95	0.89
3.	Spares and subassemblies for long term maintenance	1.30	1.22
4.	Maintenance management and labour	10	8.48
	(A) Direct maintenance cost (1 + 2 + 3 = A)	20	16.96
	(B) Cost of downtime (2 × A)		
	Total maintenance cost (%) (finished steel) inclusive of lost opportunity (A + B)	30	25.44
	Potential saving		4.70

The maintenance cost can be estimated by knowing the following:

- Number of breakdowns with their levels.
- Downtimes of the equipment for want of repairs/maintenance.
- Consumption of spare parts and other related material.
- Penalty cost, if any, due to non-availability of equipment to the users.
- Money spent in additional maintenance facilities.

The task of estimating the maintenance costs is difficult, since it depends on the frequency of failures and efficiency of maintenance services. Operating personnel are responsible for equipment usage but the maintenance personnel shoulder the responsibility for the maintenance function. The maintenance manager has to deal with this dual problem tactfully for an effective planning and proper control of maintenance cost. Analysis of maintenance cost is very essential, prior to making any decision regarding replacement of a machine or any of its components. The cooperation of production personnel can bring down the number of breakdowns and therefore, the maintenance cost.

IMPACT OF MAINTENANCE COST

The cost of maintenance may differ from one organization to another depending upon the importance attached to the maintenance function and its implementation. It is seen that for critical equipment the cost of maintenance is normally high because in such cases, preventive maintenance is followed. The maintenance cost includes cost of spares, materials, personnel and expenditure incurred on supporting facilities such as electricity, water, air gas, etc.

The other dimension of maintenance cost is the downtime of the equipment. Practical experiences reveal that the downtime cost can be much higher than the actual maintenance cost. It may be even double the cost of maintenance for some specific type of equipment/ services depending upon its utility and requirement.

The demand, for the product being manufactured by the company plays an important role in so far as the downtime cost is concerned. To maximize the production, the availability of the equipment has to be high. The downtime of the equipment can be minimized through planned maintenance and by increasing the life of the components and subassemblies of the equipment. This can also be achieved by monitoring the condition of the equipment at appropriate time intervals. An improved component life means less frequent replacement and reduction in workforce requirement for such jobs. All these factors help to minimize the overall cost of the maintenance function. A planned maintenance system can provide the maintenance function at minimum cost with better confidence of the operating personnel.

The maintenance cost comprises of two factors - the fixed cost and the variable cost. The fixed cost includes the cost of supporting facilities including maintenance staff, and variable cost accounts for the consumption of spare parts and other facilities required during the time of maintenance requirements. Thus, it is seen that the variable maintenance cost has direct relationship with the maintenance function and reducing the maintenance activities or utilizing the spare parts of improved quality can minimize this cost of the component.

The maintenance cost does not have any direct relationship with the production cost. It may be minimal during peak production months and very high during low production periods. This may be because of management's deliberate policy to defer the maintenance activity either due to fund constraints or production requirements. Even painting and denting may help in increasing the life of the equipment, but these are normally deferred due to lack of funds or production targets.

PRODUCTION COST VS MAINTENANCE COST

As indicated above, the maintenance cost can be controlled using developed maintenance practices/materials. However, the cost of production is the function of raw material cost, processing cost and of course operating and maintenance cost of the equipment. To produce quality product, the cost of raw material and processing cost cannot be compromised and once again the cost of operating and maintaining the equipment/system can be minimized. Therefore, the cost of maintenance plays a vital role in controlling the production cost. The other way to control the production cost is to follow an improved processing technique, which may save sometime during machining/processing.

The concept of total productive maintenance discussed earlier can also be helpful in minimizing both, i.e. production cost as well as maintenance cost. Here, both groups work in liaison with each other and the life of the equipment/system is enhanced due to their mutual cooperation and understanding. The optimum use of the machinery is also possible since the objectives of both the groups are the very same to produce a product at minimum cost and earn maximum profit for the organization.

MAINTENANCE BUDGET

For effective working of the maintenance department it is essential to set aside certain amount of money to meet the maintenance expenditure, which is termed maintenance budget. The maintenance budget can be of the following types:

- Appropriation budget, which sets aside money for each activity independently depending upon the nature and requirement of work.
- Fixed budget, which means fixed allocation is made for a specified period of time to meet out all expenditure related to maintenance functions.
- Variable budget, which allows expenditure based on the maintenance activities, as and when they arise.

For efficient working of the maintenance department, it is desirable that all the above types of budgets be judiciously prepared. The biggest challenge in preparing such budgets is the unpredictability of maintenance jobs. It is really difficult to forecast the cost of breakdowns as the level of such breakdowns it is not known and the time and material requirements cannot be decided.

For the purpose of cost control, it is relatively easy to implement material and time standards for production operations. However, the same is not possible in the case of maintenance function. The decisions to be taken in the case of maintenance function depend upon various factors, which are fluctuating in nature. To aid decision-making, comparisons can be made with various existing systems after obtaining the cost data.

The following factors should be considered during the process of decision-making:

Preventive Maintenance vs Breakdown Maintenance

The relevant information regarding failure data, the downtime cost arising from breakdowns, cost of damage, loss of production, penalty cost and cost of replacement of damaged parts, and so forth should be analysed to compare the nature of maintenance to be adopted under a given situation. It will be advisable to follow breakdown maintenance, since the preventive maintenance will be costly and time consuming. At the same time for other equipment/systems only preventive maintenance can be recommended, depending upon their requirements.

Replacement vs Repair

If the cost of maintenance as well as that of operation is increasing, it is desirable to replace the affected equipment/part. Here, the benefits from the replacement must be known to adopt the particular policy, however, the cost of equipment/service must be kept in mind. It should be the general policy of the management to take care of deterioration and obsolescence before deciding for any of the above policies.

Hired Maintenance vs Departmental Services

The variable as well as fixed elements of cost of the maintenance department can be collected and a comparison made when the same facilities are hired from outside. In the present context it is seen that the hiring of the services can be more effective and economical in the case of public enterprises because of high fixed costs that include manpower and material costs. The liability of the organization is also minimized, as they are not held responsible for any untoward happenings/accidents, etc.

Overhaul Periods

Schedules can be prepared to repair the equipment during its normal downtime in order to reduce the production cost as well as the machine downtime cost. Here also, relevant costs such as the cost of production, hiring charges and downtime cost can be collected for the purpose of comparison. Some of the maintenance work will affect the performance of machinery, can be carried out at the schedule time period to control the maintenance costs.

Maintenance in Time vs Time Available Maintenance

In some places, it is a normal practice to run partially failed equipment for the purpose of meeting the production targets. This may deteriorate the condition of the equipment further. Information regarding cost of loss of efficiency, damage to equipment, etc. can be collected for a better decision. However, such practices must be discouraged in order to be able to achieve the satisfactory performance of equipment/services.

It is possible to decide about a particular policy, if the cost of relevant maintenance and related functions are known in advance so that better maintenance cost control can be exercised.

Cost of Obsolescence

To match the productivity of newly available machines, the older machines are often put under excessive stress. This results in higher expenses in fuel, oil, wages and maintenance of older machines. Such costs are accounted as cost of obsolescence. If the old machines are not thus utilized, then also they suffer from cost of losing opportunity. Sometimes running of obsolete machines cause excessive downtime due to non-availability of essential spares in time since the production of such equipment/system have been stopped by the concerned organization.

Cost Reports

These reports must be prepared by the maintenance manager with the help of the accounts department to know the followings:

- Comparison of actual expenditure against the budget
- Information about the variations in costs with details
- Maintenance cost reports to intimate expenditure actually incurred
- Total project costs if manned by maintenance department

MAINTENANCE COST CONTROL

Maintenance department has to exercise effective cost control, to carry out the maintenance functions in a pre-specified budget, which is possible only through the following measures:

- First line supervisors must be apprised of the cost information of the various materials so that the objective of the management can be met without extra expenditure on maintenance functions.
- A monthly review of the budget provisions and expenditures actually incurred in respect of each center/shop will provide guidelines to the departmental head to exercise better cost control.
- The total expenditure to be incurred can be uniformly spread over the year for better budgetary control. However, the same may not be true in all cases particularly where overhauling of equipment has to be carried out due to unforeseen breakdowns. Some budgetary provisions must be set aside, to meet out unforeseen exigencies.

- The controllable elements of cost such as manpower cost and material cost can be discussed with the concerned personnel, which may help in reducing the total cost of maintenance. Emphasis should be given to reduce the overhead expenditures, as other expenditures cannot be compromised.

It is observed through studies that the manpower cost is normally fixed, but the same may increase due to overtime cost. However, the material cost, which is the prime factor in maintenance cost, can be reduced by timely inspections designed, to detect failures. If the inspection is carried out as per schedule, the total failure of parts may be avoided, which otherwise would increase the maintenance cost. The proper handling of the equipment by the operators also reduces the frequency of repair and material requirements. Operators, who check their equipment regularly and use it within the operating limits, can help avoid many unwanted repairs. In the same way a good record of equipment failures/maintenance would indicate the nature of failures, which can then be corrected even permanently.

A timely submission of equipment for service/repair by the production personnel can contribute towards reducing the rate of material consumption, because it would avoid further deterioration of the parts, which can be repaired and therefore do not require replacement. The objective of cost control is to achieve savings. For this purpose, estimations of economic life, total cost of owning, running and maintaining a machine are essential. As shown in the Fig. 17.2, if the average life cycle cost of a new machine is known and the combined cost of operating and maintenance of an old machine is evaluated, the exact time for replacement can be decided and savings of some expenses can be obtained.

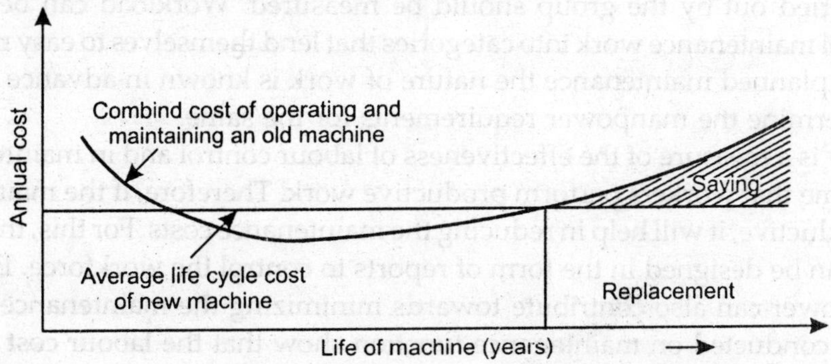

Fig. 17.2: Cost based Replacement of Machines

OPTIMIZATION OF MAINTENANCE COST

As discussed in the above paragraphs, the cost of maintenance has various factors, which include, cost of material men and overheads. Certain components of this cost are constant and no compromise can be made, however, the variable components of the cost can be optimized through various means. One way could be, to hire the services of the qualified and trained persons who can complete the maintenance work in time and therefore, minimize the maintenance cost. The other way could be to purchase good quality equipment/system, which will have less maintenance problems. It is seen in practice that goods from a particular company are more popular than the other, because of their quality and reliability. Initial cost of such equipment may be high, but their operating cost and maintenance cost may be low, which can compensate for high initial cost in due course of time.

Maintenance policy being followed in the organization will also have its affect on total maintenance cost. Organizational objectives will have their impact on the maintenance cost as well. The parameters discussed under maintenance budget should be given due consideration to decide about the feasible system which can be followed in a particular organization.

ABSENTEEISM AND OVERTIME

One important factor leading to higher maintenance cost is the cost due to absenteeism and overtime. Because of fixed manpower provided for the maintenance work, the cost of labour is influenced by absenteeism of the labour and overtime payments made to them. Even though workers may be absent due to valid reasons such as medical leave, earned leave, etc. being regular employees, other persons may have to be hired on ad-hoc basis and this ultimately increases the cost of maintenance. Sometimes the regular employees are paid overtime, thus increasing the maintenance budget. To minimize such extra payments, the maintenance department needs to follow strict policies to combat absenteeism and overtime. A study can be conducted to frame a policy concerning the maintenance workforce composition and size. The workload should be measured. This will help to ensure the smallest and most effective workforce. By knowing the workload, the number of men required to complete a job can be determined. For every group of maintenance workforce, the targets must be set and then the actual work carried out by the group should be measured. Workload can be measured by dividing the total maintenance work into categories that lend themselves to easy measurements. For example, in planned maintenance the nature of work is known in advance and therefore, it is easy to determine the manpower requirements for the same.

Productivity is a measure of the effectiveness of labour control and in maintenance it is the percentage of time that workers perform productive work. Therefore, if the maintenance work is viewed as productive, it will help in reducing the maintenance costs. For this, the performance measurement can be designed in the form of reports to control the workforce. Effective use of available manpower can also contribute towards minimizing the maintenance costs.

The studies conducted on maintenance function show that the labour cost can be a good measure of labour efficiency. It has further been observed that by upgrading the planned maintenance activities through proper planning and scheduling, it is possible to substantially reduce the labour cost.

Effective planning of maintenance work is one of the most important ways to minimize the maintenance costs. Planning ensures that material, labour, tools, work place, etc. are properly arranged and utilized. This helps the maintenance crew to complete the job in a well-organized and productive manner. It is seen in practice that a good amount of time is wasted by maintenance personnel in laying their hands on the above facilities, and such infructuous effort be minimized through better planning and control. The supervision of maintenance work is also very essential to control the costs. Many times the work is undertaken without proper approval and planning, which not only results in improper use of the maintenance workforce, but also leads to improper distribution of maintenance workload.

COST OF ACCIDENTS

It is seen in practice, that none of the maintenance systems offer continuous running of machinery without any failure. However, some failures can be predicted and others can be minimized. Accidents are the part of the system and can not be avoided, though minimized. During such accidents the cost of maintenance may go very high and therefore, arrangements must be made in the maintenance budget to meet out such eventualities. Provisions are to make the equipment/system operational again, after accident. Sometimes, the accident is very severe and the recovering of equipment may not be economically viable and hence replacement may be needed. Funds must be arranged for this, to maintain the services of the equipment or machinery. Such funds can be managed through insurance if this is being practiced in the organization.

QUESTIONS

1. Discuss the importance of evaluating maintenance costs.
2. Explain how maintenance costs can be kept under control.
3. What is the impact of maintenance cost on production costs? Discuss.
4. Highlight the significance of maintenance budget.
5. Discuss the methods used for controlling maintenance costs.
6. How maintenance costs can be optimized? Explain.
7. Discuss the importance of absenteeism and overtime in maintenance organizations.
8. What are the life cycle costs of a plant? Discuss in detail the operation costs of a plant.

18

Maintenance and Environment

INTRODUCTION

Environment is nature made, but the mankind is disturbing it, for their own selfish gains. The equipments/machines designed, and manufactured are to be used in the same environment and therefore, the impact of environment cannot be overlooked. The functioning of any machinery is obviously influenced by the working environment, and the degree of maintenance and maintenance performance will therefore, be governed by the working conditions. Wear and lubrications are of the primary concern of the maintenance personnel. These two aspects depend on load, speed and working temperature. The selection of lubricants is affected by the prevailing environmental conditions. The working environment also affects the type of maintenance to be adopted. Therefore, maintenance practices vary from place to place, from industry to industry and even from location to location of the equipment in the same organization. The success of a maintenance system is largely dependent on the working conditions, which vary from place to place. The same piece of equipment may manifest quite different types of maintenance problems, depending of the site conditions.

The adverse effect of environment on machines in petroleum, fertilizer, coal and mining industries is more prominent because of hostile working conditions. The main source of pollutants in these industries is dust and gases, which affect all metallic and non-metallic parts of the machinery. For example, it is seen in chemical and mining industries that the impellers of the centrifugal pumps need to be replaced frequently if they have to handle acidic water or abrasive slurry. The maintenance cost of such machinery increases due to adverse working conditions, some of which can be improved if the working personnel give proper attention. For supporting moving shafts bearings are used in almost all the machines. The main causes of bearing failures are dust and improper lubrication. Therefore, the lubrication system of a machine often needs more attention of the maintenance personnel. The influence of the environmental factors on lubrication system needs due care in a greater detail to minimize the problems. For this purpose, the reference of the following factors will be necessary.

- Technical requirements of lubrication
- Contamination of lubricants
- Operational factors
- Temperature and humidity.

Each of the above factors is discussed in detail in the following paragraphs:

TECHNICAL ASPECTS

Lubricants are applied to reduce wear and the frictional forces among the matting parts. The type of lubricant and the manner of its application are normally decided, based on the load and speed of the machinery. Sometimes under special conditions a lubricant also works as a coolant and thus, if forms a part of the environment. Under such circumstances it is either mixed with the fuel or circulated in the system. Loads and speeds may be fluctuating, uniform or intermittent in nature or may be a combination of all these. The present trend is to design compact equipment capable of withstanding high loads and speeds. In case of hydrodynamic bearings, the lubricant used has to form a thin layer to separate out the two matting parts, which will reduce friction and heat in the bearings. So, the role of lubricants has become more prominent and crucial which calls for their better quality.

The quality of lubricants is also of prime importance and it deteriorates faster under ambient conditions. Care has to be exercised in maintaining the quality of lubricants by proper storage and handling. Since these are lived items, users must pay proper attention to use them before their expiry dates. Storage of the lubricants must be made at a place where proper ventilation, light etc. are provided. The good lubricants can withstand the adverse effect arising due to presence of water in the lubricant. Certain oils sometimes may cause problems by forming a stable emulsion within a short interval of time. In such situations, a change in the viscosity of oil occurs and this causes operational problems. Water effluence, therefore, is an environmental factor in the lubrication system and must be taken care off.

Water used as coolant may affect the system if it is not of proper quality. The quality of water is dependent on the extent of the dissolved materials, which varies with the nature and place of its source. Acidic elements in water may damage the whole system resulting increase in the frequency of failures. Such water, when used in systems where it remains stagnant for sometime may cause corrosion problems. In the area of opencast mining, the use of dump trucks is very common. As observed in these areas, that the failure of radiator is more common in summer season as compared to winter season. This is primarily due to the use of bad quality of water by the operators. The leakage of water into the lubrication system can also cause severe problems, which must be taken care of.

In steel and thermal power plants, under certain processes the temperature of the system rises very high, which may affect the quality of lubricants being used there. Proper quality of the lubricants must be used at such places and due attention be given for their periodic replacement so that the systems under use are not affected adversely.

ENVIRONMENTAL IMPACT ON LUBRICANTS

Contamination is the process, where lubricants loose their required properties and render unfit for use. These lubricants reduce the life of the equipment/system if used continuously without their timely replacement. The nature of contaminants will largely depend on the nature of products/materials etc. handled during plant operation and also working environment. This process starts from the lubricant store itself. Improper handling of lubricants will be the main cause of their contamination at the store level. If containers are not properly covered, the atmospheric dust is likely to enter the lubricant directly. The contaminants may be classified as under:

- Solids, such as dirt, dust, and abrasive material, metal wear particles, cotton fluff, rubber cuttings, etc.
- Liquids, which can be soluble and insoluble acidic bodies being products of oxidation and improper combustion of fuels.
- Miscible liquids in the case of water contamination in bearings, or water and fuel dilution in internal combustion engines, which may cause serious problems.
- Some lubricants get oxidized when it is in contact with free air, which is very common in any working environment.
- Additives used in oil may cause contamination in a long run.
- Emission of gases and their reaction with the lubricants can also contaminate them.
- Degradation due to plant or engine deterioration.

For permissible contaminant concentration, no limit can be set in advance. A machine can be taken out of service as a consequence of high concentration level, must be based on the source, diagnosis, the concentration trend and relevance of deterioration. The concentration level of metallic contaminants depends on their origin (major or minor alloying element), the type of equipment and its operating characteristics and nature of defect. In certain cases, it is observed that the equipment is operational in spite of high copper and iron concentrations. In other cases a small amount of contaminant may cause failure. However, the manufacturers for concentration level based on the experiments are providing some guidelines and these must be followed to avoid further deterioration of the equipment/system in use.

Fuel oil dilution in an equipment/system causes a decrease in viscosity, whereas oxidation and presence of insoluble causes increase in viscosity. Flash point of the fuel also changes due to change in viscosity, which itself is the function of temperature change. Some of the insoluble do not have any appreciable solvent action on the product, but a rapid increase can have serious implications.

Suspension of insoluble water due to condensation of products of combustion due to any reason (defective gaskets/cracks on the cylinder head etc.) will promote formation of sludge and, therefore, may interfere with the despersive properties of the additives, which must be taken care of by the maintenance personnel.

To overcome the problems caused by contaminants, the lubricants must counteract the ill effects of contaminants. In practice, oil/grease must be periodically checked to assess the remaining life, which is normally specified by the manufacturers. As the working environment may contaminate the lubricants, the periodic assessment of its life at regular intervals is desirable. Attempts, therefore, be made to minimize the degree of contaminants for a particular application so that quality of lubricants do not change. Poor quality lubricants accelerate the deterioration of machine parts and assembly.

The various contaminants monitoring techniques used include SOAP and magnet plug inspection analysis. Ferro graph and X-ray spectrometry are of interest for the analysis. Adding a solvent to a specimen, centrifuging to remove clear layer from the top by evaporating the solvent, weighing the insoluble, can carry out oil degradation analysis, which remain present in lubricants.

Any foreign solid substance present in lubricants will work as abrasive and will accelerate the wear process of the parts/assemblies. These can be found in bearings, which will reduce its normal life. These can also be noticed in hydraulic oil used in systems as a result of in-built particles, which enter system, contaminating particles generated by the system and particle arising out of reactions within the system. To monitor these, the oil is to be withdrawn from the system, filtered and compared visually on a microscope with a master slide.

The core/earth faults in transformers can produce localized very high temperatures which slowly breakdown the oil and produce ethylene gas. Further, gas-in-oil analysis can be used to detect gas-evolving faults. The insulating materials used in electrical systems may face the problems of gases, which arise from electric discharge or thermal breakdowns. These gases may take very long time to diffuse out of the material. However, gas-in solid monitoring can monitor the same.

OPERATION FACTOR

The quality of lubricants, also depend on the operational factors of the equipment/system. The following are the main factors, which must be cared during the use of the machinery:

Process Impact

In many cases the speed of machines, and the total load taken by it are the important parameters to decide, the type and grade of lubricant that can be used. Any deliberate change in such parameters will affect the quality of the lubricants in use and therefore its life as well. The quality of lubricants in general and viscosity in particular is solely responsible for maintaining a correct thickness of film that will keep two mating parts away to minimize excessive generation of heat between the parts, to avoid seizure. In the process of hot rolling in steel plants certain parts/components are subjected to high temperature, the lubricants, which can sustain such high temperatures, must be used to minimize failures. The engines which are frequently started and stopped, there is possibility of more carbon deposits inside the cylinder, which may necessitate frequent changes of lube-oils as a part of maintenance schedule. It is observed in practice that viscosity will change throughout the various parts of an engine according to local temperature of the parts.

Working Conditions

The very purpose of selecting good quality lubricants is defeated in the actual working conditions are not taken into consideration. This in particular relates to roads, for vehicles, ambient conditions for machinery and environmental conditions for highly specialized equipment. The dust from the roads will deteriorate the lubricant at a much faster rate than without it. The local temperature will also affect the quality of the lubricants. Humidity and moisture in the atmospheric air will affect the working of air-conditioning systems and ultimately the quality of lubricants used therein. The emissions of toxicant gases from the internal combustion engines on the road will also affect the lubricants in a long run through chemical reaction.

Skills of Operator

The skill of operator is of a paramount importance in handling the equipment/system. In moving vehicles, sudden braking will cause excessive overheating of the parts, close to brake systems, leading to deterioration of lubricants at faster rate. This also increases loading on the engine as well, which is also likely to get overheated and again will affect lubrication oil used there. Technical knowledge about the equipment/system handled by the operator, is also necessary to minimize excessive utilization and therefore, reduction in failure frequency. Less skilled workers do attempt to carry out maintenance related functions, which may work against the interest of equipment, as per the saying "Little Knowledge is a dangerous thing". Everybody around tries to impose his opinion without having actual knowledge of the subject, which may further complicate the situation and may create more problems than their solutions.

The skilled and trained operators can take care of routine maintenance works, which are looked after by the maintenance personnel in normal course of duties. The other advantage of this system will be, reduction in failure of the machinery with regular check up by the operators. The operator, who is made responsible for running a particular machine, will definitely care for it more than the other persons. If the equipment is not operated and its normal working conditions can also create more maintenance related problems, once again due to the mistake or deliberate attitude of the operator.

Fire Hazard

The petroleum products are always a source of fire hazards and a lot of loss is sometimes incurred owing to the lapses on the part of maintenance personnel in up-keeping such items. Selection of lubricants/oil etc. should be made according to their actual place of use. During operation, many of the components/parts are subjected to high temperature; the lube-oils used there must be fire resistant. The other reason of fire could be leakage of oils/lubricants through joints/sealing, wherever possible, the concept of temperature control may be applied to protect the lubricants as well as equipment/systems.

TEMPERATURE AND HUMIDITY

The selection of lubricants should be carried out with proper consideration of temperature and humidity of the working environment. The environmental factors such as temperature, humidity, light and so forth affect the quality of the lubricants even during their storage. Therefore, lubricants storage must be designed considering the prevailing temperature, humidity and light conditions. To maintain the quality of lubricants during their storage, it is essential that they must be properly kept in containers duly covered with lids. Their withdrawal from containers should be done carefully without mixing with each other. The transportation of lubricants for use on machinery should be done under protection, to avoid open exposure of sun, light and air, which will contaminate it.

The store in-charge must be a trained person to maintain the quality of the lubricants besides looking after the handling and storage facilities. The replacement of lubricant must be in a planned manner and the containers must not be left uncovered so that the undesired influence of the environmental condition is reduced. In India, weather conditions change from place to place with temperature varying from one extreme to another. This, environmental condition is also responsible for high cost of maintenance of some of the equipment used under such conditions. The quality and grade of lubricant suitable at one place may not suit the other and due attention must be given to this factor.

Besides temperature, humidity i.e. percentage of water contents in the atmospheric air, has significant effect on the equipment/system. It is observed in coastal areas that the machineries are more prone to corrosion due to presence of moisture and salt. Special treatment will be required to protect the machinery/system and also to minimize the failure frequencies. The life of the equipment used in such areas is also reduced considerably due to environmental factors as compared to that operating under normal working conditions. Therefore, environmental factors play an important role in the selection and up-keep of the equipment/machines. If the two are not matched properly, it will increase the maintenance cost. Therefore, proper discretion should be exercised while purchasing the equipment/system keeping in mind the adverse effect of environment.

DESIGN IMPROVEMENTS

As discussed in the earlier paragraphs, the design of the equipment/machines can be improved to overcome the adverse effects of the terrain conditions under which the system has to operate. If the temperature rise is due to the speed of the rotating parts, it can be kept within the specified limit to save the system. For the same purpose, many of the machines are provided with alarm systems. When the preset parameters reach the highest limit, in certain cases the system will get shutdown automatically to avoid the sudden breakdown.

Multipoint fuel injection system (MPFIS) used in the recently developed internal combustion engines is another example of design improvement, where the emission of toxic gases from the engine is minimized. The basic concept in this system is the complete combustion of the fuels, which was not possible in the earlier systems. This also incorporates fuel economy, easy start of engine and pollution control.

The starting of cold internal combustion engine is very difficult, because very little amount of fuel gets vapourized initially and therefore a rich mixture is required. This problem becomes more acute in cold regions where temperature remains below sub zero. The cylinders of such engines are specially designed with the provision to heat them from external source to achieve easy start. Here, heating is done with the help of heaters, bulbs to raise the temperature of the cylinder and make vaporization of the fuel easy.

Material selection for the machinery used under adverse environmental conditions also calls for proper attention. Under such conditions if the failure frequency is high, it immediately needs a thorough investigation to pin point the failure cause. By simply changing the material the problem can be overcome for example, impeller of a centrifugal pump used for pumping slurry and other such materials.

The thermal efficiency of the internal combustion engines largely depends on their cooling system also. Therefore, some design changes in cooling systems can bring substantial improvements in their performance. At the same time less amount of heat will be thrown in the atmosphere, which again adversely affects the systems around it. These days some additives are being used with cooling water to improve thermal radiation. For the same purpose, the exhaust valves of some of the engines are specially designed and filled with sodium for effective cooling. Due to over-heating of internal combustion engines, the requirement of cooling water will also go high.

At cold places and in winter season, the use of anti-freezers becomes essential if the engine remains out of operation for sometime. The water in the jackets, radiation tanks and leading pipes will freeze, expand and lead to their fractures. These additives lower the freezing point of the water.

The concept of design should be such that minimizes the maintenance work and incorporates maintainability. The more reliable system is, the less maintenance, it will require. Modular design of software can improve its maintainability.

ENVIRONMENTAL FRIENDLY SYSTEMS

To minimize the adverse affects either on the environment or from the environment, of the existing machinery/system in use, the concept of eco-friendly systems is gaining momentum. The emission of hazardous gases from the moving vehicles or from factories are causing severe burden on the environment. Pollution of any kind let it be air, water or sound, will leave its ill effects on human beings as well as the equipment/machinery around it. Protection of nature is the responsibility of each and every one of us to save nature.

The use of electric trains for transporting men and material is one step towards eco-friendly systems. In big metro cities, where pollution level is very high, attempts are being made to run the buses by using C N G gas, which emits less harmful toxicants. In the same way at some places battery driven vehicles are being used to minimize pollution. The use of oil and gas fired boilers is very common to replace coal fired boilers. Government of India has banned and abolished the use of locomotives being used by the Indian Railways.

The introduction of multi-point injection systems in internal combustion engines used in cars and vehicles is mainly to minimize the pollution and make the vehicles eco-friendly. The use of light indicators on the vehicles provided for easy travel on the roads, with minimum usage of horns/sirens to minimize the sound pollution. For the same purpose, better silencers are also being used to emit minimum sound in the environment.

Pollutants from the chemical industries, human waste etc. are carried away to rivers, canals and to ponds. The water at such places becomes unsafe for human beings as well as animals. The introduction of water treatment plants, helps not only in removing the harmful pollutants but also makes the water re-usable. The seepage of discharged water into the earth also spoils the natural water available there because of harmful ingradients.

In the opencast projects, a huge amount of dust is generated due to various operations of digging and transportation. This dust is harmful for human being as well the equipment being used there. By sprinkling water the suppression of dust is the normal practice to create healthy environment. Growing of more and more plants and deforestation of the area is also an essential requirement imposed by the regulatory bodies.

The re-processing of the solid waste can also minimize the pollution related problems. One of the ways adopted in this area is the use of Gobar gas plants. In our country, particularly in rural areas, cow-dungs are used as firewood to cook food and other articles. The dung discharged by the animals can be used to generate gas, which can be used to cook food as well as for lighting purpose. The waste can further be used as manure in the fields. In Russia, one of Gobar gas plant, having 25,000 pigs, the power requirements are met by using the dung of the animals and, thus, saving environment as well as electricity.

CONCEPT OF CONSERVATION OF ENERGY

Friction and wear are associated with any moving part of the machine/equipment. No motion is possible without friction, but at the same time, a higher value of friction consumes more power or fuel. It is well known that when the piston rings of an internal combustion engine gets worn out the efficiency of engine goes down and it consumes more fuel. Wear is the integral component of friction in any system. High frictional resistance is further associated with rise in system temperature and increased wear rate. All the basic elements of machines such as bearings, gears, valves, cylinder and piston require lubrication to minimize friction and wear. Implementing preventive maintenance and replacing worn-out parts before they fail, can also make the system's/machinery's energy efficient.

Design Factors for Energy Conservation

Due to high cost of petroleum products, it is desirable to select fuel-efficient equipment/ machines. The earlier concept of going in, for over design for the sake of safety was at the cost of high consumption of fuel energy. With the development of new techniques it has been possible to implement the design without causing any risk to the safety of the equipment or personnel.

To save energy in a hydraulic system for example, attempt should be made to control the temperature as effectively as possible to minimize the loss of energy due to overheating of components/parts. In certain cases where the load is supported on hydraulic pressure, the running of the pump over long time periods may cause excessive heat, but the energy can be saved using accumulator, or a variable volume pumps. The variable displacement pump can be used with appropriate controls to minimize power losses. The hydraulic circuits should be designed to save energy. One example of it is the use of solenoid valve in place of ordinary relief valve. These valves are normally used when fluid is supplied to several systems at different maximum working pressures.

While selecting drives for operating machines, it is important to know the actual power requirements. It is generally seen that designers select over size drives, firstly for the sake of safety and secondly for future expansion work. High power drives consume more fuel or electricity, which can be minimized through proper selection of their size.

Due to increased pace of mechanization the demand for electricity power has become excessively high and the need of the hour is to pay serious attention to the problem of electrical energy conservation. It is, therefore, required to identify the areas where it is possible to conserve electrical energy. One effective measure can be the planning of proper sites for the sub-station transformers to minimize distribution losses. The proper selection of energy efficient drives can also be a step towards minimization of losses in the overall system. It is important to use high efficiency machines for a continuous operation; even-though initial cost of the machines may be high. The high cost can get compensated in due course of time by way of improved utilization/efficiency.

The other way to conserve energy can be through the improvement of energy efficiency by implementation of the following short-term measures:

- The unnecessary idle running of electric motor and equipment is minimized.
- The induction motor should not run at full load as their efficiency decreases with decrease in output load under fluctuating load conditions, hydraulic motors provide better efficiency.
- The capacities of the equipment working together in system must be matched so that all equipment operates at their rated capacities.
- Proper use of compressed air can also sometimes lead to significant energy savings. The air losses in the pipelines and valve joints should be reduced as far as practicable. Moisture in air does not directly hamper energy savings, but moist air causes leaks and thus damages equipment/system.
- The transformers must be placed near the load , to decrease the transmission losses.

- Selection of materials for fabrication of parts/assemblies in case of automobiles should be given due care since overall weight of machine/system consume more power. The modern automobiles are energy efficient on this account only.
- Use of high efficiency systems should be given preference over the others.

Long Term Energy Conservation Methods

These measures include major equipment design changes or implementation of new techniques, which not only give direct energy savings but also provide cost effective utilization of equipment.

It is the general view of the experts that energy consumption in industrial sector in our country can be reduced by 20% if proper measures are taken. In this regard training of personnel also plays a very significant role towards improvement of the performance of the equipment/machinery. Skilled and trained persons can be asset to organization in attaining defined goals in the area of energy conservation.

Besides the above, the following can also be considered as the means for the conservation of energy in their respective fields:

Thermal Plants: For production of electricity, one of the main sources is the thermal power plant where, working fluid is steam. To improve the efficiency of thermal power plants, the quality of the steam has to be improved, which can be done with the help of economizer, pre-heater, evaporators and super-heaters. To improve their performance further, the following methods can be used:

(*a*) Use of binary vapour cycle, since single fluid does not meet all the working requirements.
(*b*) Use of coupled cycles where two Rankin cycles have been used in series.
(*c*) Use of combined cycle plants; higher energy conservation efficiency from fuel to electricity can be achieved, since the combined power plants operate through a higher temperature range. The combined plants may be of the following types:
  - Gas turbine — steam turbine plant
  - MHD — steam plant
  - Thermionic — Steam plant
  - Thermoelectric — steam plant

It is observed that with the application of above methods, a substantial amount of energy saving can be achieved with reduction in fuel consumption.

Automobiles: The multi-point fuel injection (MPFI) system has been successfully installed in automobiles to achieve fuel economy. The main advantages of this system are improved combustion of fuel, uniform distribution, quick throttle response, easy cold starting high volumetric efficiency and minimization of fuel condensation in the chamber.

In some of the diesel engines, turbo-chargers are used to improve the volumetric efficiency of the engine, which helps in better utilization of injected fuel, therefore its conservation.

For automobiles, the road conditions play important role in fuel consumption. In case of poor roads, the fuel consumption is always high since vehicle cannot be run on high gear ratio.

Energy Management: With shrinking fuel availability, the need to conserve natural resource is of paramount importance. The following methods of energy management must therefore be adopted:

- To meet the peak load demand, interconnect power networks that might have different demands on them.
- Use the most efficient power plants for base load power generation and the old, less efficient plants to meet power generation requirements during peak loads.
- Install the smaller low capital cost power plants for example, gas turbine and hydroelectric plants.

Incorporate energy storage systems, which can store surplus energy and can be released at the time of need. This is a new concept but will have applications in future.

QUESTIONS

1. Discuss the adverse effects of environment on industrial equipment.
2. What is the impact of environment on lubricants? Discuss.
3. Explain the operational factors that will influence equipment/systems.
4. How temperature and humidity affect the system? Explain.
5. Discuss how design improvements can bring out environmental friendly systems.
6. Explain the concept of conservation of energy and its impact on environment.

19

Maintenance Material Planning and Control

INTRODUCTION

It is often seen in practice that many of the equipment/services remain unavailable for utilization for want of spare parts. More often than not, such a state is usually due to poor planning and control of spare part inventories. The importance of proper spare parts management cannot therefore, be overemphasized for effective use of maintenance manpower and support systems. It is also known by experience that most of the breakdowns are repetitive in nature. Therefore planning of spare parts is much easy. During the purchase of equipment, a huge amount of money is spent on spare parts inventory, which can be better utilized through effective management. The basic issues attached to the spare parts management are:

- High financial involvement for the procurement of spare parts.
- Overstocking of spare parts due to unpredictable nature of failures as well as to reduce the downtime of the equipment/facilities.
- Prolonged usage of equipment beyond its normal life cycle due to scarcity of funds also increases the cost of spare parts requirements.
- Non-availability of proper type of spare parts in the local market due to obsolescence of equipment, as a result of advancement and improvement of technology.
- Non-interchangeability of spare parts due to different types of models, and specifications pose problems to the users.
- Purchase of imported equipment also causes a lot of problems in procuring spare parts in a reasonable time particularly for the lived components.
- Political relations also affect the supply of spare parts in the case of critical in-service equipment procured from foreign countries by defense forces and several other industries.
- Government's import policy may also affect the supply of imported items.
- Provisioning of funds for stocking costly items may be a constraint.
- Communication gap between the concerned parties, i.e. suppliers, users and intermediate agencies, may also cause some problems for the proper stocking of inventories.
- In India, machineries are used for much longer period and the manufacturers do stop production of such products and getting spare parts is very difficult.

PRESENT MATERIAL MANAGEMENT POLICIES

Through studies, it is known that most of the organizations in the country today do not have proper warehousing facilities for storing the items. For example, if the rubber items need to be stored in an airtight compartment, their storing elsewhere will affect their life and the equipment will not give proper service.

The technology transfer from the developed nations also causes the problem of interchangeability of items since the same are procured from different countries, although for the same type of industry.

Many organizations set up overhauling facilities, which require procuring of the (spare parts) in advance since the date of overhauling is known. But for the first overhaul, the estimation of exact number of spare parts is normally difficult. Sometimes, if a particular item is not available in the spare parts inventory, it has to be locally manufactured/fabricated. This also causes delay in completing the overhaul of the equipment.

In developed nations due to fast pace of development and induction of new technology, the production of spare parts for the superseded equipment is not continued for long, since it is taken for granted that no spare parts would be needed and a new version of equipment will be produced in due course of time. In developed countries, the normal practice is to discard an item rather than to repair it, as the repair cost is sometimes much higher than the cost of the new item because of labour/production cost. However, during the manufacture of the original equipment, sufficient quantities of spare parts are produced, which are normally supplied during procurement of equipment, but the main problem remains with lived items which cannot be stored for long time period.

Most organizations follow A B C analysis for procurement of fast moving inventory items, but for slow moving and costly items, criticality analysis must be carried out. The concept of economic order quantities used by some organizations is sometimes purely of academic interest and does not serve the purpose in the desired manner. The overhead expenditure involved in the procurement of spare parts, which has an important role in the preparation of maintenance budget, is also not known to many organizations.

For standardized equipment the concept of spare parts bank is ideal and is being followed in many industrial organizations. The spare parts kept in a warehouse at a central place are supplied to the concerned units on demand. This also reduces the inventory cost, as each unit in the organization does not have to store the same items. The following points need due consideration while in planning stage.

- The concept of substitution of imported spare parts is also growing due to their general non-availability. This also helps in saving the foreign exchange. There are some difficulties being faced in establishing the total substitution of equipment/items for the various reasons such as:
- Unwillingness of the concerned people with the fear of mission failure.
- Non-availability of the complete specification data, and material requirements.
- Lack of proper production facilities and materials availability.
- Higher cost of indigenous items at initial stage due to rigorous testing.
- Contract conditions from the suppliers at the time of procurement of equipment/system.
- Manufactures entail uneconomical quantities, which put financial burden on the organizations.

INFORMATION SYSTEM

An information system for spare parts, to furnish relevant information to the indenter, if available at the user level, can help in procurement of spare parts at the appropriate time and at minimum cost. Properly maintained logbooks are very useful to know the past history of the equipment, which will also help to know the type and correct quantity of particular spare part to be ordered. It is seen that the maximum portion of maintenance budget is consumed in procuring the spare parts. Keeping in view such problems, it is advisable to develop a spare part information system, to keep information about equipment/system updated. These days the use of computers can help in a big way to maintain the correct information pertaining to material requirement at each level to take a correct decision.

CLASSIFICATION OF SPARES

Every organization has its own way of classifying the spare parts, depending upon its requirements, into various categories such as critical spares, fast moving spares, slow moving spares, consumable and non-consumable spares etc. However, such classification should be based on the type of maintenance that has to be carried out, such as overhauling or preventive maintenance. Maintenance spares are those, which are required during routine maintenance work, which include seals, belts, fuses, bearings, etc. Overhauling spares are those, which are required in a lot at the time of complete overhaul of the equipment. The requirement of each type of spare depends upon the failure frequency and time interval of overhauls. Their quantities may be different depending upon the maintenance policy being practiced.

Role of Spare Parts

The spare parts are required to replace a defective item, whose performance has fallen below the specified level. Storage of spare parts play an essential and preventive role in any organization because they make it possible:

- To get right amount of material at the required time of need, to ensure continuous, uninterrupted working.
- To avoid physical impossibility and economical impracticability for getting right quantity at right time.
- To facilitate economic working of maintenance function because of its supply in time.
- To make effective use of available capital and/or storage space to avoid over storing or under storing.

INVENTORY COSTS

It is important to bear in mind, all the cost elements associated with an inventory during its procurement and storage, because it consumes a good amount of maintenance budget as well as time. The following main elements of an inventory costs should be taken care of:

1. Capital cost involved in procuring the items.
2. Recurring storing cost involved in warehousing the items.
3. Loss due to non-usage of stored items, theft, and wastage obsolescence of equipment due to change in technology.

4. Overhead expenditures in terms of payment of wages, rent, etc.
5. Insurance costs to protect the stored items against fire, theft and other related risks.
6. In the case of computerized inventory, payments made towards the computer charges.
7. Stock out charges incurred on account of emergency procurements at the time of actual maintenance work.
8. Penalty costs paid for not supplying a particular product in time if due to production halts on account of breakdowns, which could not be repaired because of non-availability of spare parts in time.

SPARE PARTS COST OPTIMIZATION

The spare parts cost optimization is possible through many well known techniques available, which include proper forecasting of inventory items, scientific analysis of maintenance activities by failure analysis and other methods to predict failure time and to get the information about spare parts requirements. Inventory control, refinement of purchase procedures to reduce the lead-time, spare parts consumption control and standardization of products is also possible. The actual need of a particular spare part arises when it is required for replacement. If the same is available in stock the machine downtime is minimized. The objective of an efficient inventory should be to keep the equipment downtime as low as possible; this will, in turn, also optimize the spare parts cost.

As indicated earlier, A B C analysis was in practice for inventory control along with VEIN-VED (vital, essential, important, normal - vital, essential, desirable) analysis. Looking into the limitations of the above concept, MUSIC-3D (multi unit spare inventory control-3 dimensional) approach came in existence. And it was realized that MUSIC-3D had numerous advantages, as compared to others to reduce the spare parts cost and therefore being used in almost all the industrial units.

It is of paramount importance that the spare parts supplied should be of high quality and reliability. The need of reliability becomes more pertinent in the case of high precision equipment mainly used in defense forces and nuclear power plants where risk factor is very high.

Before actually accepting the spare parts, inspection must be carried out to see that the parts conform to the desired specifications with respect to accuracy and reliability. This will also reduce the chance of defective items being stored in inventory. The procedure of inspection will, however, depend upon the type of equipment and testing facilities available at the source organization. All items cannot be tested and have to be accepted principally.

Factors Affecting Inventory

The main problem associated with the inventory is to answer questions such as how much to order? And when to order? These can be answered by using inventory models. However, the main factors are demand and cost. Many factors related to above two factors include:

- *Economic parameters:* Here the purchase cost/manufacture cost of component/part, its procurement costs, holding costs, stock-out costs etc. are important for consideration.
- The demand of the spare parts can be deterministic or problematic. In case of deterministic it is easy to calculate its requirements but if difficult in case of problematic nature since failure pattern is unknown.

- *Ordering cycle:* It is identified by the time period between the successive placement of orders, which can be done with continuous or periodic review of the existing inventory.
- *Lead-time:* It is the time period between the placement of the requisition for an item and its receipt for actual use in the shop. It can be deterministic or problematic in nature depending upon supplies.
- *Time horizon:* It is the time period for which inventory is to be controlled, which can be finite or infinite. In most cases it is done on yearly basis.
- *Number of supplier:* There can be several stocking points in the inventory system starting from warehouse to retailer and finally to the customers.
- *Number of items:* Inventory involves more than one commodity and as the number of item increases, the space requirements will also change accordingly.
- *Availability of items:* The most important parameters of inventory are its availability. If the item is available, other parameters can be managed. When the same is not available, alternatives are to be thought off, so that the maintenance work does not suffer.

SPARE PARTS PROVISIONING

The success of the maintenance department depends upon the timely availability of the right quantity and quality of spare parts. If proper record of maintenance activities is maintained, it will be possible to know the consumption pattern of spare parts for working out the provisioning quantities of the same for use in the future.

The basic objective of a maintenance function is to eliminate random failures or reduce their occurrence. If the preventive maintenance system is followed, then it is essential to know the number of components that need to be replaced during the maintenance. Such quantities can be determined in advance for making arrangements for their procurement. However, in the case of random failures, it is difficult to predict the exact number and type of spare parts required, but by experience it can be known with greater accuracy. The tendency of maintenance staff to store maximum number of spare parts should be discouraged as it involves high costs. The number and type of spare parts required to be provisioned, would depend upon the following factors:

1. Criticality of the spare part, which will make the equipment totally inoperative or unsafe for use. This is a very important factor and should be given due attention.
2. Lead time required for procurement of each spare part.
3. Lived components, which have stipulated life in terms of running hours or have life on the basis of calendar period. Such parts must be changed after expiry of their life period otherwise they may cause damage to equipment/system. No lived component should remain in use after expiry of its life in the interest of safety of the equipment/system.
4. Cost is the primary factor for determining the actual type and the number of components/ spare parts to be procured. It is unsafe to keep on reconditioning the equipment without caring for its material life. The costs saved in this manner may be harmful for the equipment/ system.

For sophisticated and complex equipment the spare parts management is much more difficult. The problem further complicates in the case of imported equipment. To provision the right type of spare part in right quantity and at the right time requires considerable planning in terms of quality, quantity, price, material, and other parameters of delivery transportation, etc.

When a spare part is to be ordered, the first thing to be finalized is its correct specification, which can be done only after having the complete details about equipment/system. If the item is procured from the original equipment manufacturer, then reference to correct part number can give the relevant information about the component/item. However, the possibilities of indigenous substitution must be explored in respect of imported spares, many more manufacturing details need to be made available or may have to be developed through trials, for satisfactory production of the substituted part.

Due attention should also be focused for the improvement of spare parts to improve the efficiency and life of the equipment. All relevant information available about a particular spare part should be passed on to the manufacture for possible improvements in design and fabrication. For some of the items, suppliers may not agree to manufacture the item due to uneconomical quantity, and then the same can be developed at the organizational level or through local manufacturers.

For smooth working of an organization, formal systems are essential which also applies in the case of provisioning of spare parts. A proper system of ordering must be developed to reduce the lead-time and paperwork. The merits and demerits of existing systems must be examined for effective improvements.

STORE MANAGEMENT

The inventory of spare parts has to be stored in a warehouse. The management of the warehouse is an important function since it acts as a buffer between the user and the manufacturers. The main functions of managing a warehouse include:

- Receipt of orders
- Processing of orders
- Procurement of spare parts
- Inspection of spare parts
- Issue of spare parts
- Stock record keeping
- Valuation of stock
- Inventory control
- Stock verification at appropriate time
- Proper storing of spare parts
- Coding of spare parts
- Coordination with other departments/agencies
- Sorting out store problems

As large number of activities are thus associated with store management. On the other hand, there are various factors, which influence the working of stores, such as warehouse location, availability of proper ventilation including air-conditioning, facilities installed for shifting heavy material, and procedures followed for preservation and safety of stores. Due care of lived items should be taken to avoid their wastage. Measures to protect stores against fire, corrosion, dust and attack by insects, moisture, and sunlight, etc. should be duly implemented.

Store management has to be applied across the board from the receipts of items to its proper delivery to the users. During the planning of the input process, the store team has to perform various functions such as:

1. Determination of proper quantities for provisioning
2. Preparation of requisitions for provisioning
3. Procurements of stores
4. Quality assurance
5. Coordination of all relevant work including documentation.

The above activities have to be systemized so that overcrowding of items is minimized. The items being received in the store must be checked thoroughly for their quality, quantity, and expiry date of the parts, if any.

The output function of spare parts includes its issue to different departments or sales to the customers. The lived items should be issued on first come first served basis. Whenever, the stock of items falls below the pre-specified level, the procurement action must be initiated. Sometimes if an item is not available in stock, then the purchase department must initiate the required formalities on operational basis, so that the item can be made available to the user in time. Proper control of the indent quantity and issue will therefore help to minimize the storing cost as well as, will provide better information about future requirements.

The concept of multiple warehousing can also be implemented to cover large areas with an established distribution system. This practice is normally followed for public utility items and is more advantageous for costly items. These items can be stored in a central place and can be supplied at the time of need. But for such a system, a well-established information system is desirable. Adopting this system can also minimize the cost of procurement and inventory costs since items are stored at central place without its duplication at different stock points.

The objectives of a company dealing with the sales of spare parts must be to ensure availability of right quality of spares at the right time to the users with minimum investment at the seller's end. The creditability of the supplier will depend upon pricing of spares, efficient handling of customer orders, and reliable time delivery with sound after sales service.

The important function of any organization is to keep proper and adequate quantity of spare parts in its stocks to carry out maintenance work efficiently in the stipulated time period and at minimum cost.

SPARE PARTS INVENTORY

For the purpose of maintenance work, the availability of spare parts in time is a must; otherwise the whole work might come to a standstill. With this in mind, the concept of inventory becomes essential for any organization having a number of equipment. The main problem that arises is that of inventory or spares control, that is, how much to stock in stores, as every time procurement, involves money and time. Where the equipment is imported, the problem of inventory becomes more acute as good amount of time is involved in procurement of the spare parts.

Normally every manufacturer of equipment provides certain items as spare during the sale of equipment and also recommends which items should be stored and in what quantity. Therefore, it is the task of the maintenance group to properly estimate the number of spares required for the work. Various models are available which can be used to estimate the required number of spares. However, in the case of bigger organizations spread over the entire country, a number of stores are required at different places with central store at the headquarters.

Forecasting the requirement of spare parts is the most difficult task for the maintenance department. However, the demand for spares depends on the following factors:

Failure Rate

It is important to know the failure rate of the components/parts over a specified period of time. This would help to know the number of spare parts required. The requirement of the spare parts also depends upon the following factors:

- Time needed to procure the items
- Availability of spare parts in stock
- Number of equipment needing spare parts.

Any effective material management system is based on the accurate forecast of material demanded. In the case of maintenance work, this forecast is critical as the occurrence of failures is unknown and unpredictable. To actually find the requirement, the theory of probability of failures can be applied which will indicate the frequency of failures. If the population of equipment being used is high, this assessment will yield more accurate results otherwise it becomes hypothetical.

In the computation of spares requirement, two factors to be considered are:

- The cost of downtime of equipment/service
- The cost of stocking spare parts
- A balance between the above two factors has to be struck to minimize the inventory costs. In certain cases due to non-availability of items a particular item may have to be stored on one-time basis though its use may be rare since the same may not be readily available or its failure will make the equipment out of use.

Spares Control

For the success of the maintenance function, the following important factors need to be looked into:

- Provision of accurate quantity of spare parts
- Reduction in the non-moving spare parts, which add to inventory cost.
- Maintaining the optimum level of spare parts inventory so that the maintenance work does not suffer.

In practice, it is difficult to strike a proper balance between the above three factors, therefore, some controls are necessary for the same. The following are a few steps to minimize the range and scale of spares:

- Efficient indenting of spares
- Control over consumption
- Recondition/overhauling of used spares
- Use of management techniques
- Effective involvement of maintenance/operational personnel.

During the indenting process, the obsolescence of equipment and experience of usage of a particular system, must be borne in mind. During schedule maintenance work, effective control over consumption of spares must be exercised. The parts to be replaced must be checked before their replacement. Sometimes it so happens that even-though a particular subassembly has been changed during the operation of the equipment because of sudden failure, the same becomes due during scheduled/planned maintenance work. This must be properly checked and avoided. At the same time, efforts must be made to reuse the item after reconditioning, but safety aspects of the system must be given due consideration.

In case of maintenance the demand for spare parts cannot be known with certainty and therefore, use of multi period problematic inventory models will yield better inventory control. These systems require periodic review of the inventory and where no model meets the requirement, Monte Carlo simulation can be used for effective control of the inventory.

QUESTIONS

1. Why maintenance material planning and control is required in any organization? Discuss.
2. How information system can be useful in planning maintenance material? Explain.
3. Discuss the classification of stores based on the need of the organization.
4. Explain how spare parts cost can be optimized.
5. Highlight the importance of spare parts provisioning in maintenance organizations.
6. Elaborate the steps required in store management.
7. Discuss how requirements of spare can be controlled.

20

Manpower in Maintenance

INTRODUCTION

Manpower is the most valuable and unique resource for any industry. The performance of manpower is unpredictable, because of human behaviour, which is not so in the case of other resources, such as machines, which are man made. Unlike other resources, labour is capable of exercising individual judgment and freely determining the course of its actions depending upon a particular situation. Therefore, manpower planning is one of the most critical jobs in an industry. Such planning needs to consider the industrial interests with special responsibility in a humanistic sense.

Manpower is the key economic resource, which should be paid the due attention of the management as is being given to others such as, finance, equipment, raw materials, production, sales, investments, profits and maintenance functions. A company cannot hope to forecast its future manpower requirements accurately, unless these are related to future production, sales and maintenance levels.

The rationale for manpower planning is related to the efficiency and effectiveness of the organization. Manpower planning is a process, wherein the demand and supply of labour are matched for effective utilization. Profit maximization is often a goal of an industry. To achieve this goal, capital and labour must be managed properly. Based on the price of labour and nature of the work, the industry must determine the requirement of manpower for the overall life of the industry. It is essential to find out what type of manpower will be recruited and at what time during the life of the industry. These aspects are the important considerations in manpower planning.

The need for manpower planning arises from many facets of the organization. It is unwise to assume that skilled workers will be freely available in the labour market, whenever required, and so an adequate period of time, must be set aside to organize and run appropriate programs to meet out such requirements. With the workforce being maintained at an optimum size, with the proper mix of skills and experience, problems resulting from staff shortage or surpluses, on account of production stoppages, promotion beyond real capabilities, inefficiencies and staff dissatisfaction may be alleviated.

However, it is not claimed that the planning process will always provide exact forecasts, and this is rarely achieved in practice since human beings display a commendable resistance to change. Rather, it is suggested that there should be a more realistic expectation for the process to give general indications of the potential trouble spots. Such indications are invariably based on assumptions; these being implicit in whatever forecasting procedure are used. Simulations of the possible outcomes arising from different assumptions about the factors and variables under study should provide a real aid to the manager. The inevitable slight inaccuracy in the forecast should not be taken as an indication that the process is a waste of time. Instead, it serves to emphasize the need for manpower planning to be a continuous process, in which forecasts are checked against reality, and models and assumptions are constantly evaluated and updated to accommodate the varying nature of jobs.

The first step in the manpower planning process is to establish the planning horizon, i.e. the future period for which the manpower will apply. The aim being that the demand for labour services should remain matched to the supply of labour services over this period. The demand for labour is estimated. This can be done by considering the technical relationship between the input of labour services to a production process and the output of the commodities generated by the production process.

Based upon the above information the current manpower stock is established, and estimates are made of the likely wastage or labour turnover during the forecast period. This wastage may then be subtracted from the current stocks to give an estimate of the supply of labour in the forecast period, and if necessary, an allowance may be made for normal recruitment during the period. The difference between the estimates of demand and supply of labour services is often referred to as the manpower gap, and one of the main components of manpower strategy is to formulate plans for closing the gap perhaps by recruitment and training or by planned redundancy. The period over which forecasts are made will inevitably vary from one organization to another organization. Depending on such factors, as the training time required for new employees, time period over which new capital equipment is developed and installed and the stability of the environment in which the organization operates will be factors to be kept in mind before deciding the manpower forecast program.

The models used in planning should explicitly contain variables relating to the rate at which labour is utilized, factors determining wastage or characteristics of the turnover process, and factors influencing the supply and demand for labour. Manpower planning also needs to consider inputs like the learning process and the characteristics of the system of industrial relations. Economics, sociology, psychology and other applied sciences all have a part to play in a good manpower model.

OBJECTIVE OF MANPOWER PLANNING

A Manpower planner's function in a maintenance system as part of an overall planning team is:

- To interpret maintenance forecasts, in terms of manpower requirements for a particular time period, with the emphasis that the maintenance jobs are normally unpredictable.
- To indicate manpower constraints as a result of company's policy for the future expansion as well as any other changes required for addition and deletion of products.
- For maintenance functions, the forecast for manpower requirements are different as compared to other jobs, therefore due care must be taken while following any policy and model for it's planning.

The acute shortages, particularly of skilled technical and qualified manpower in many companies, have brought home the need to anticipate manpower requirements a few years ahead of their actual work assignment, if they are to ensure an adequate supply of qualified manpower. It is again important to see that manpower resources are utilized as effectively as possible. Manpower planning may be conceived as a strategy for the acquisition, utilization, improvement and preservation of the company's manpower resources.

STAGES OF MANPOWER PLANNING

Manpower planning for maintenance needs, should be carried out following a systematic methodology to minimize the problems associated with it. Such planning involves the following stages:

- Defining the overall maintenance objectives for the stated time period in advance.
- Converting these maintenance objectives into manpower objectives for the same period, allowing for changes in the maintenance system, number of equipment, and so on.
- Designing a manpower information system to obtain and process data in the most efficient and economical way.
- Undertaking a manpower inventory in terms of composition, deployment and utilization by tabulation and careful analysis of the current manpower resources within the system.
- Analyzing manpower requirements or the demand forecast based upon the maintenance objectives. This requires:
    - A decision concerning as to, where the maintenance system or the workshop management should concentrate its efforts.
    - An assessment of the overall size of the maintenance system keeping in mind the possible changes if any.
    - Estimates of the resources needed by the system to meet out the maintenance requirements effectively.
    - Formulation of advance maintenance plans so that the other related arrangements can be taken care off.
    - Detailed targets for each management level to achieve their targets as planned.

To forecast the relationship, between the workload of the maintenance plan and the manpower required. An analysis of the past manpower performance and trends in the adoption of maintenance practices is needed. The forecast must also provide not only the type of manpower required but also the category/level of the workers.

ANALYSIS OF MANPOWER SUPPLY

A vital part of the manpower planner's work is to indicate the constraints on policy imposed by conditions in the local and occasionally national labour market. The planner needs to consider the internal supply of labour already existing in the organization. Labour turnover also affects the internal supply of manpower as do promotions and transfers and consequently, these too will have to be analyzed for adequate manpower planning.

IMPROVING MANPOWER UTILIZATION

After hiring or procuring manpower, the management must consider the ways and means of improving the quality of manpower utilization. Manpower utilization is measured in terms of achieving targets with improved productivity. Consequently, the determination of efficiency indicators and measures of productivity enhancement must be incorporated in the manpower management system. When these have been established, then shop management can set targets that are achievable and can be monitored. It is possible to indicate the achievable targets by comparing the present performance with the trends in the past in the same shop floor or at other places of work.

In the light of these, the recruitment policies must be examined and developed. This helps in formulating policies required to improve manpower management particularly with respect to absenteeism, illness, accidents, and disputes.

Training Needs

The development of manpower and the training scheme should be planned on the basis of the future needs for trained as well as untrained manpower. Consequently, the first step is to identify the training needs within the maintenance system, and the next step is to assess the effectiveness of such training policies and methods. Both these jobs are closely dependent on the nature of the industry and must be carried out following a methodology suitable for the situation. The training needs basically arise due to procurement of new equipment/system and the effectiveness can be judged with the performance evaluation of the concerned worker.

CONTROLLING MANPOWER COSTS

A systematic appraisal of manpower costs gives, a better appreciation of the contribution and value of various manpower activities such as recruitment, training, and development. It also quantitatively appraises such problems as absenteeism, sickness, and labour turnover. This will assist in making cost benefit analysis and in deciding on the approach, which will give the best subsequent yields.

Time Scale of Manpower Planning

Manpower plans are basically of two types—short-range and long-range. Short-range manpower plans are detailed and from them, stem factual plans for recruitment, promotions, transfers, training, etc. Short-range plans, however, are much more concerned with the strategic problems and indicate the overall direction of manpower policies. As such their use lies in their value to shop management in deciding the maintenance policy.

Consequently, the lines of approach are different for short-range and long-range manpower planning. For long-range plans, the manpower planner, on the basis of analysis, has to translate the maintenance objectives into a forecast. He can then discuss such a forecast with shop management in order to incorporate their assessment of the situation.

However, with short-range plans, the emphasis is reversed in that the shop managers make their own forecasts based on their intimate knowledge of the situation. This forecasting is to be supported by the data and analysis of the manpower planner. The short-range plans must be sufficiently flexible to cater to changing circumstances.

Forecasts of both demand and supply may turn out to be wide off the mark, and the main assumptions on which they are based may require substantial revisions. To make provisions for these, two procedures are normally followed. First, the manpower planner must closely monitor the analysis and forecast by comparing them with what actually happens on the ground. This should provide an early warning of the sort of adjustments necessary and at the same time, enable the manpower planner to test and improve the analysis and forecasting methods that were used. Secondly, there should be some system of reviewing the manpower development by workshop management at regular intervals. The following factors are important in the contemplation of manpower planning in any organization.

DEMAND AND SUPPLY OF LABOUR

The technical relationship between the input of labour services to a production process and the output of commodities generated by the production process is depicted by the production function. Variation in the flow of output is partially determined in the production function by various inflows of input in labour services over a particular period of time. Man-hours worked, should be used as an indicator of labour input since it takes into account the variations in the number of hours worked per employee and thus more accurately reflects the flow of labour services.

The problems of labour supply are multidimensional. They concern the decisions of individuals and form the financial and social constraints faced by both the individuals and the organizations. The individual is the cornerstone of supply analysis, as individual decisions in aggregate create the supply stock, which confronts and interacts with the job offers of firms.

Age Distribution of Workforce

The age distribution of workforce is important to ensure that the age structure is not unbalanced in a particular sector or category of employees. It might be discovered, for example, that there are too many employees of the same age in a particular department or occupation and steps have to be taken to avoid promotional difficulties, which may arise as a result. The age pattern is also useful to determine the effect of a policy of enforced retirement at any selected age for any particular trade or department and also to determine the cost implications of any such policies.

Length of Service

This gives a breakdown of the labour forces by period of service. The particular value of a periodic analysis of this nature is that it highlights the departments or occupations in which it is difficult to get the labour to settle. It will also show the size of the hard core loyal to the company and thus, help to improve the effectiveness of recruitment procedures and devise measures to curb labour turnover.

Geographical Locations

An analysis of, where employees live is useful when considering the location of new work sites or possible movement of labour. If such developments are long term, an age limit should be established so that older workers due to retirement in the intervening period are not included in the analysis.

Labour Turnover

The extent of voluntary and involuntary terminations in a company is important to know not only for purpose of manpower supply, but also as an indication of the general health of the organization. Compilation of the wastage statistics, even if done thoroughly, is of little value unless the figures are interpreted to discover the real cause of labour turnover and action taken accordingly to control wastage.

Analysis of labour turnover must be carried out in detail. The total wastage rate alone will not indicate whether or not a situation is satisfactory. Even if the total number of persons leaving an organization is by itself not considered excessive, a more detailed breakdown may reveal the areas that are unsatisfactory. A statistical breakdown of labour turnover by each department in itself may indicate the source of the problem.

It is also important to identify the type of employees in wastage analysis to see whether or not a lot of key people in any function are being lost. A further breakdown by grade will show at what level people are being lost. What action should be taken to equalize these losses will be, for the personnel department to decide, depending on just how large the talent drain is. If it can be shown that the company is consistently losing talent over a period of years, it is best to start with the rectification of the company recruitment policy.

Reasons for Leaving

An analysis of the reasons given for termination will indicate the steps needed to reduce turnover and its effects and whether or not the cost of any measures to reduce turnover can be justified. It is likely that some simple measures can remedy a costly situation. Furthermore, such an analysis will indicate ways to improve management-employees relationship as well as job effectiveness. The reasons for leaving can be broken down under the following headings:

- Economic factors which affect the labour market and impinge on the work situation such as the level of employment.
- Factors within the organizations, which affect the employees in terms of promotion/ demotion.
- Factors concerning the employee as an individual or factors affecting a group of employees.
- Factors related to the working environment/working conditions which may influence the employees.

The reasons for leaving will vary from one type of individual to another. Opportunities for promotion, work culture, lack of independence and interruptions in the workflow, size of the work group are some of the reasons given by people for leaving the organization.

Analysis of External Supply

If a company is considering an expansion of its workforce, then together with an analysis of the internal supply situation, the manpower planner will have to analyse the external manpower supply as well. This helps to find out whether or not enough manpower with the required skill is available locally or whether a different labour market would need to be accessed.

MANPOWER FOR MAINTENANCE SYSTEMS

The factors that must be considered while planning manpower for maintenance systems are enumerated here, taking the example of mining industry. However, this can also be applied to other organizations with suitable modifications.

Type of Operation

The manpower planning of a mine maintenance system will vary depending on whether it is an underground mining or an open pit mining operation. Generally, the number of equipment employed in the latter case is higher. Opencast mining deploys a large number of technologically advanced machinery. Therefore, the maintenance system must be well planed with the required level of skills and the requisite number of such skilled maintenance personnel.

Moreover, the maintenance site should also be properly selected for easy transportation of the faulty equipment. Based on the conditions of the working site and the type of machinery used, specialist manpower would need to be developed.

Extent of Mechanization

The extent of mechanization of the concerned mine will decide the manpower strength of the maintenance system. The mechanization level will dictate the skill of the maintenance personnel to be used in the workshop. Special training needs are associated with complex mechanization. In case of a particular mine management plans for a modernization program, motivating and training the existing maintenance manpower for new challenges of maintenance operations may have to be looked into very carefully.

Type of Mined Mineral

The type of ore or mineral has a bearing on deciding the strength of the maintenance manpower. The working conditions in a mine are often adverse and very challenging, which result in frequent breakdowns demanding more care from the maintenance personnel. These aspects must also be considered in deciding the manpower requirement in a mine maintenance system.

Number of Shifts

The number of maintenance personnel required depends on the number of shifts practiced. A three-shift working needs more maintenance manpower than a double-shift working. In a three-shift working, operation-crew is also trained to cope with emergencies and first-aid maintenance work.

Size of Mine

The size of the mine is determined by the targeted output capacity. This is one of the most prominent factors directly related to the strength of the manpower required for mine maintenance. The more the output, the more is the number of machines employed for production. This invariably requires more maintenance personnel as well.

System of Mine Maintenance

The manpower requirement for maintenance also depends on the system of maintenance being followed. The manpower requirement is different for preventive maintenance, breakdown maintenance and condition-based maintenance. There is difference in the level of sophistication or skill required on the part of the maintenance personnel under different circumstances. The strength of manpower required will be the highest for preventive maintenance and the lowest for the breakdown maintenance for a particular type of equipment.

Strategic Location

The strategic location of the mine in relation to manpower supply also affects the manpower strength. If the availability of skilled personnel is low in the adjoining area then the requisite manpower may also have to enhance recruited from outside. Under such a situation, the manpower turnover has to be studied carefully in order to understand the factors, which would help to minimize the turnover.

MAINTENANCE MANPOWER ESTIMATION MODEL

In practical situation, maintenance planner for not making the machines available for preventive maintenance even though meticulous care is taken for scheduling the equipment scientifically experiences difficulty. This non-adhering to the schedule poses a cumulative problem to maintenance organization as a whole. Because of the fluctuating workload, which is sometimes time dependent, planning of maintenance workforce is problematic. Knowing the total number of equipment coming in for various types of maintenance work, a scheme for determining the number of machines, fitter, skilled worker and helper, etc. can be evolved by considering two main components of costs namely, the cost of manpower and cost of equipment hiring.

In practice, however, any planned schedule is subjected to change because of various constraints, which cannot be anticipated in a pre-planning stage. In such a situation it is always preferable to provide a marginal allowance in a schedule. The amount of allowance is mostly dependent on past experience and the working environment of the organization. In industry where equipment have been deployed at various places, an equipment due for maintenance may not be released for scheduled maintenance will join the spillover. This leads to a dynamic situation and due care is being taken for some of the irregularities which are the part of the system because of its inherent features.

Assumptions for Model Formulation

The arrival pattern of the equipment is known with the type of crew required with their skill and number. Considering fixed number of days per week, schedules can be drawn for the arrival pattern of the machines to the workshop. If the work on any particular equipment is not completed in a week's time, it is then taken up in the next week. Therefore, model is developed with varying percentage of spillover. There can be three situations:

- When manpower is idle
- When equipment is idle
- When manpower meets the equipment requirements.

Keeping the cost parameters, optimum number of men for the work is calculated week-wise and also taking care of spillover as well. It is seen from the model that crew size can be calculated week-wise with minimum cost.

Very often maintenance planning cannot surmount for the difficulty of the repeated re-scheduling, and the main reason for such inappropriateness is non-availability of spare parts. However, especially with the knowledge of optimum spill over allowed, one can always question, if the spillage of the scheduled equipment is more than optimum one and thereby knowledge of non-availability of spares can be brought to the notice, if that is the one of the main reason of the delay, can be used to plan effectively.

As the maintenance load in any maintenance workshop is not static throughout a month, the workforce is subjected to hiring and firing by management especially for casual and temporary hands. The resource allocation is possible month-wise.

MANPOWER ESTIMATION IN CONTINUOUS PRODUCTION SYSTEM

Here a assembly line of continuous manufacturing item was chosen for study. The approach included the following.

Once the operation and work stations for different production volume are identified, norms, minimum number of workers of individual grade for respective operations are collected from norms or sequence charts of time studies. Man days for individual operation for certain increased production are calculated and tabulated for same grade of workers. The total man days for a particular type of worker are found by adding the same for different operation in the system. Total labour cost was found from the individual and categorized requirements of workers. The wage for different categories of workers was assumed at an average rate fixed by the management.

Worker grade	*Man days required for normal production*	
	Theoretical	*Actual*
A	5.02	7.0
B	47.17	55.0
C	1.23	2.0
D	15.96	19.0

The workforce estimated and expressed in man days for a particular change in percentage of production may have a high idle time. In that case, idle man days are utilized by a bit increased or decreased production and for that, some more workers may be needed or reduced at the operation.

EFFECTIVE MANPOWER UTILIZATION

As the nature of maintenance work is unpredictable, proper utilization of the manpower is essential. It can be improved by following the suggestions, enlisted below:

- Manpower planning must be recognized as an integral part of maintenance planning. The planner must understand needs and objective of the workshop.
- Support from the top management is essential for manpower planning at all stages.
- Manpower planning responsibilities should be centralized and coordinated at all levels.
- Recording of information should be correct and complete to give future guidance.
- Planning period must be long enough to take care of unforeseen problems faced during the planned period.
- The planning techniques used should match the available data and degree of accuracy required.

- Data collection, analysis, planning techniques and the plans themselves, need to be constantly revised and improved in the light of experience gained.
- Utilization of the manpower can be more effective, when the equipment due for maintenance is made available to them in time. The other support facilities must also be provided to ease the working.

QUESTIONS

1. Discuss the need and importance of manpower requirements for maintenance functions.
2. State the basic objectives of manpower planning.
3. What are stages of the manpower planning? Discuss.
4. What is meant by time scale manpower planning? Elaborate.
5. How manpower planning for maintenance is different from normal manpower planning?
6. Discuss the models used for manpower planning.

21

Maintenance Equipment and Facilities

INTRODUCTION

For successful completion of any activity, particularly depending upon external support, it is essential that the same be implemented at proper place and time. The maintenance function depends upon many external factors and prompt availability of its services plays an important role in effective completion of the jobs. The major maintenance facilities include well-equipped engineering workshops and a store organization. Sometimes loading and unloading facilities are also required. Depending upon the type of maintenance organization adopted, these facilities are provided both at the unit level and at corporate level. Generally, small workshops and stores are provided at each unit to provide daily requirements of maintenance function and central workshops are provided for specialized jobs. However, there are various advantages of centralized facilities, which include:

- Cost effective utilization of a maintenance manpower and equipment/facilities.
- Better spare parts management with the help of central stores.
- Quality work with the association of available experts in the field of maintenance.
- Standardization of work, which is repetitive in nature and is being carried at many places.
- Reduction in maintenance/repair time by utilizing, modern maintenance practices.
- Scope for better research and development activities to improve all above parameters.

On the other hand, the greatest requirement of the centralized system is that it should be an efficient and effective organization with proper coordination and control of the manpower and other facilities being employed for the maintenance work.

WORKSHOPS

Since the workshops are the essential parts of the maintenance organization, their structures as well as the equipment located therein, largely depend upon the type of machines to be maintained and the type of maintenance systems being practiced. However, the general layout of the workshops should be more or less of similar type to meet out the basic requirements. The basic objectives of a good workshop are the following:

- It must have all the basic facilities of production and fabrication for effective utilization of equipment used in plants.

- It must have all the facilities for carrying out minor modifications to the components/parts of the equipment.
- It must have provisions to ensure quality of raw materials (i.e. testing facilities) to be used for maintenance work.
- It must have a research and development wing for devising new methods of test and repair, particularly for imported items.
- It must have the requisite facilities for training the maintenance personnel working in the field.
- It must maintain a proper record system for the utilization of manpower and equipment at its disposal.

Sometimes it is not possible to have all the above facilities in the workshop for maintaining the equipment. For the major tasks of overhaul and preventive maintenance, the facilities not set up due to financial constraints can be subcontracted from original equipment manufacturers or their specialized persons can be called to the site to help in major repairs and overhauls.

Workshop Classification

Depending upon the nature of maintenance requirements and the organizational set-up, workshops can be divided into various groups such as:

- Central workshop
- Unit workshop
- Apprentice workshop
- Auxiliary workshop
- Mobile workshop/trains
- Commercial workshop
- Special workshop

The central workshop is normally well equipped with modern test and repair facilities. It usually has a big machine shop with different types of lathe machines, milling machines, drilling machines, grinders and other ancillary machines. In the central workshop, specialized types of jobs are taken for manufacturing or modification. The central workshop also has the facilities like welding, painting and even heat treatment. Small research and development groups are also attached to such workshops to carry out the development work, particularly manufacture of imported parts, which are difficult to get from the original manufacturers.

The unit workshops are attached to smaller establishments and have basic facilities of machining, welding, etc. to carry out routine maintenance work.

The apprentice workshops are equipped with facilities on a moderate scale for the purpose of training the technicians on the job as per requirements. This training is different from the conventional training imparted in the institutions. All basic facilities of machine shop, welding shop, carpentry shop, smithy shop electrical and instrumentation shop, battery shop, and electronic shop, etc. are made available in these workshops.

Auxiliary workshops are workshops in the conventional sense that have some facilities under temporary sheds to cater to the maintenance requirements at the site where equipment is used. This is only for saving the time to rectify minor nature of breakdowns. Trained personnel are placed at these places to deal with such eventualities. They can also be used to carry out weekly or monthly and planned maintenance program for which special shop facilities are not needed.

Mobile repair workshops are useful for attending running repairs and minor breakdowns at the site where, sometimes, it is not possible to shift the equipment to the workshop for the maintenance work. For example, if the tyre of a vehicle needs replacement due to some reason, then they have to be replaced at the site or a broken axle shaft has to be attended at the site itself. Mobile repair workshops are used in the case of railway breakdowns, or automobile failures on the road or in military operations at the forward landing areas. Moderate facilities are equipped in a vehicle or wagon to carry out routine repair works.

The commercial types of workshops are very common and are found almost everywhere. These workshops undertake different types of repair work, which includes repair of gears, pistons and cylinders of automobiles. In these shops, the damaged parts of equipment are also repaired, besides providing contract service to the users. The organizations, which do not have their own workshops, mostly depend on these commercial workshops for several types of repair work, with the help of trained persons. The work in these workshops is completed, comparatively in lesser time period.

The special workshops are established exclusively for a particular type of repair work. For example, in an automobile workshop, work related to a particular type of automobile may only be carried out. Similarly an electric shop may undertake the repair of only electric motors. These shops aim to produce quality work since trained persons are normally employed in these shops. These types of workshops are becoming very popular because of their quality work and timely service. Some manufacturers provide these types of workshops exclusively for their own products to meet out maintenance requirements.

A simple workshop organization can be of the following type:

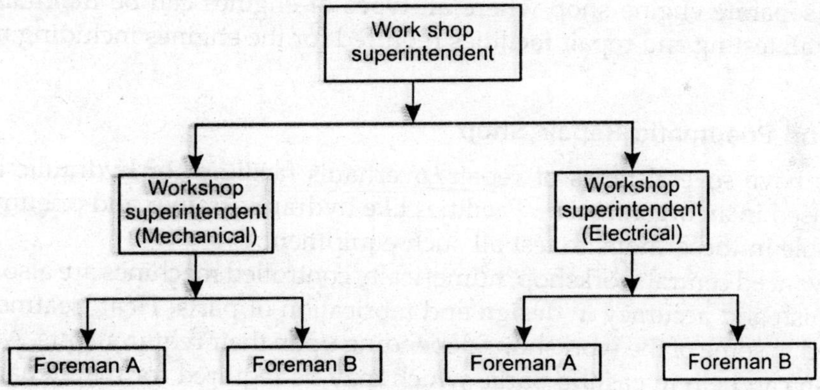

Fig. 21.1: Workshop Organization

Workshop Location

The workshop location is one of the most important parameters for the maintenance function. The selection of a workshop site should be based on the following criteria:

- Workshop should be located centrally so that all concerned departments do not have much difficulty in transporting their equipment for maintenance/repair/overhaul, etc.
- Adequate provision should be catered for future expansion and modification of the workshop to meet out the additional facilities owing to technological developments.
- The workshop is easily accessible by road and rail so that even large-size equipment can be handled without much difficulty.

- Workshop site must have the essential facilities such as electricity, water, gas and drainage to minimize the problems during their use.
- The movement of men and material should be designed to be minimum during the course of work in progress to reduce the total transportation time in the workshop.

CENTRAL WORKSHOP ORGANIZATION

The central workshop is the most important workshops for any organization. This workshop consists of a number of maintenance shops within it. The extent of facilities incorporated in these shops depends on the maintenance requirements. Mainly the type and size of the equipment to be maintained determine the size of the workshop. Normally the following shops are essentially present in any central workshop:

Machine Shop: It is one of the most important shops of a workshop, which is used not only for repair work but also for the manufacture of parts and other fabrication work. Lathe, milling machine, drilling machine, grinders, etc. are housed in this shop. Other facilities such as tool room, smithy, welding, forging and carpentry are also provided along with a machine shop.

Electrical Repair Shop: Repair and maintenance of electrical equipment can be carried out here. The testing facilities for all types of electrical equipment must also be provided in such shops. Provision for winding etc must also be provided along with other facilities.

Engine Overhaul Shop: Such shops carry out overhauling/repairing of engines. Specialized persons are normally employed in such shops. An industry deploying several vehicles usually maintains a separate engine shop where all types of engines can be maintained. This shop should have all testing and repair facilities required for the engines including non-destructive testing.

Hydraulic and Pneumatic Repair Shop

These shops have several types of repair/overhauls facilities for hydraulic and pneumatic equipment used in an organization. Facilities like hydraulic trainer and pneumatic test rig are made available in these shops to test all such equipment.

In an advanced central workshop, numerically controlled machines are also made available for better finish and accuracy in design and fabrication of parts. Heat treatment facilities are also provided in some of the workshops depending upon their requirements. A well- equipped foundry shop can help in casting parts, which may be required for maintenance functions.

LAYOUT OF WORKSHOPS

Before laying out any type of workshop, the total area required must be calculated taking into account the present maintenance load requirements of the organization and those of future expansion. The layout also depends on the type of shop, which in turn depends on the type of equipment to be maintained. The shops like plating, painting and annealing are housed away from the main shop. There are a few advantages of a centralized layout and they include:

- Better supervision and control of work,
- Good communication between different shops,

- Transportation time from one shop to another is minimized,
- Better planning and sequencing of various operations in shops are possible, and,
- Better utilization of available space and facilities.

The biggest drawback of centralized workshop layout is the interference between the shops, noise and general pollution, which may influence the working efficiency of workers. The other disadvantage can be the formation of groups, which may lead to conflicts at different levels.

The different shops must be laid in such a manner that the work progresses in a forward direction as far as possible, through the shortest possible route from the point of entry to the point of its completion. This will prevent unnecessary handling and transportation of material and spare parts required for the maintenance work. Figure. 21.2, below shows the typical layout of an engine overhaul shop. Provision for spare parts can also be made at a convenient place.

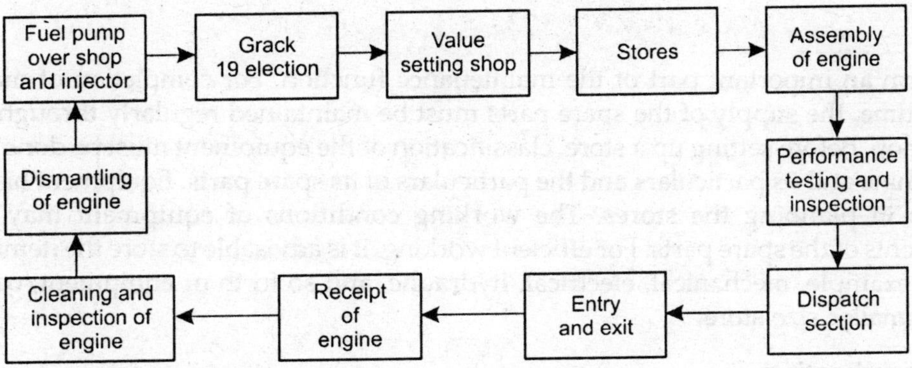

Fig. 21.2: Typical Layout of an Engine Overhaul Shop

It can be seen from the above layout that the movement of the jobs does not affect the working of the other departments/sections. Here, it is easy to control the repair/maintenance work. Wherever extra manpower is required, it should be provided to complete the job so that the sequence of operation can be maintained and the work does not get delayed.

In the receipt section, the jobs are taken over with all its accessories and requisite documentation and then passed onto the inspection shop, where the nature of the work content is investigated. The engine is then dismantled and its subsystems/parts such as fuel pump, injector etc. are sent for overhauling to different bays/shops. The engine block is thoroughly checked for any cracks in the crack detection shop. The valves are sent to the valve setting shop. Any parts, which need replacement, can be demanded from the stores. Dimensional accuracy of the components is also verified for proper working of the engine. Finally, the reassembly of the engine is undertaken in the assembly shop, after which the complete engine is tested for its performance in the testing shop. All through the quality inspectors also check the work. On completion of all work and quality checks, the tested engine is then dispatched to the concerned department/section.

Similar types of layouts can be planned for different types of components/parts such as electric motors, hydraulic motors and pumps etc. in different overhaul shops.

Workshop Control

Normally, the workshop superintendent, who is the overall in-charge, looks after the administrative control of a central workshop. It is the duty of the workshop superintendent to assign the maintenance work to the right person at the right time since he is to be held responsible for any lapses. His coordinated effort has to be geared, in improving the availability of the production/service equipment. He is directly placed under the Chief Engineer. The financial control of the workshop is implemented through its annual operational budget prepared by the workshop superintendent in consultation with his immediate subordinates and the finance manager.

To effectively utilize the resources of the workshop it is essential to plan and coordinate the maintenance work in advance. A senior foreman should carry out supervision of work in each shop. From time to time, the workers should undergo training in the related areas of maintenance and control and learn new techniques of inspection/monitoring such as non-destructive testing (NDT) and so forth.

Stores

Stores form an important part of the maintenance function. For completion of maintenance work on time, the supply of the spare parts must be maintained regularly through the stores organization. Before setting up a store, classification of the equipment must be done to identify each machine and its particulars and the particulars of its spare parts. Equipment history cards also help in planning the stores. The working conditions of equipment may affect the requirements of the spare parts. For efficient working, it is advisable to store the items category-wise, for example, mechanical, electrical, hydraulic, and so forth or equipment/components wise for smaller size store.

Store Classification

Depending upon the work requirements, the stores may be classified into various categories such as:

- Central store
- Workshop store
- Section store
- Special store

A central store is normally established to meet the overall requirement of the spare parts, where items of all the categories are stocked. It is the responsibility of the central store to procure the items as and when projected by the users. Some buffer stock of critical items must be maintained to deal with the emergency demands. Proper methodology has to be adopted for timely procurement of the items in sufficient quantities.

Workshop stores form an integral part of the central workshop and maintain stock of spare parts and materials required by the central workshop as per the production plan.

Section stores are provided to cater to the day-to-day needs of the maintenance work and therefore only small quantities of spare parts are stored in these stores. They form the auxiliary stores of a central store to minimize the routine workload of the latter.

Special stores include items such as fuel, oil, lubricants, and the like which need to be stored in a safe place away from the normal place of work.

Location of Stores

The store items are basically of two types:

(i) *Raw material:* Which can be stored even in open and unshaded place depending on its type and use.

(ii) *Finished item:* These must be stored in properly designed and fabricated bins to safeguard them from environmental effects.

The location of stores, depending on the availability of place, accessible by road/rail transport facilities, should be in close vicinity of the central workshop as far as possible. Types of store items also decide the location. For example, inflammable and explosive items have to be stored away from the domestic areas. At the same time, bulky items such as construction materials need to be stored in open place.

Normally a store should be a single-storied building, but electrical and electronic items can also be stored in multi-storied buildings for proper ventilation and provisions to protect the stored materials must be made.

Store Layout Design

The size of any store will depend upon the type of equipment and their numbers. As spare parts storing is expensive and therefore the quantity of the parts must be decided carefully through statistical analysis based on the past experience on inventory requirement of similar type of equipment. The storage area should be well planned to accommodate bulky as well as small items properly, and space should be earmarked for future expansion as well.

In designing a store layout, due consideration should be given to storing sensitive items against damage caused by moisture, light and fire. The danger of theft of small and valuable items should also be taken into account. They should be stored in a safe place. Special care must be exercised for storing the lived items so that they can be utilized well before their expiry date. Congestion of stores should be avoided to obviate the chances of damage. Similarly, if the items are to be stacked, care should be taken to avoid damage to the parts.

In case the store building has two stories, the light items must be placed in the second story and the heavy items in the ground floor. There should be proper arrangements for the movement of the stored items, i.e. stairways and lifts must be provided. For transporting the spare parts, forks, light trucks should be made use of. To store small parts of machines, bins and shelves must be provided. The bin size must match the size of items to be stored in it. Bins must have the reference number of the parts for easy recognition of the items. This will help the users to identify a particular item/component quickly.

ROLE OF LUBRICANTS FOR MAINTENANCE

Friction and wear are the most common problems associated with the industrial equipment and can be minimized with the use of lubricants. The following give the brief description of lubricants to help the maintenance function:

Lubricants are used in almost all mechanical systems and equipment and they play a very important role in maintenance engineering. The proper use of lubricants contributes towards increased life of equipment and plant and reduction in downtime and maintenance cost. At the same time, lubricants assist in ensuring the safety of operations. It has been observed that lack of proper lubrication leads to disturbing sounds emanating from the moving parts of machines. This often irritates the operator of the machine. To achieve the quality of maintenance, therefore, the maintenance management needs to take care of the following:

- To ensure that all the components/parts are provided with proper type and quality of lubricants at scheduled time intervals.
- To select the correct type of lubricant as well as the lubricating system to ensure its effective utilization.
- Periodic monitoring of the quality of the lubricants in use and their replacement within the specified time period.
- Proper storage and handling of lubricants to protect them from damage and deterioration.

LUBRICATION SYSTEMS

Lubrication systems of equipment mainly depend on the design of the equipment. They depend on the method of applying the lubricants. They also differ in their working principles. However, for the purpose of general classification, lubrication systems can be divided into the following groups:

- Manual lubrication,
- Automatic or self-designed lubricating system, and
- Built-in lubrication systems.

In the case of manual lubrication, hand grease guns and oilcans are used. The oil or grease must be filled in the cans in a non-dusty atmosphere to avoid contamination. Where grease is used as the lubricant, mostly one type of nipple should be used as far as practicable, since it will ease the greasing operation. The oilcan should also be dust proof and fitted with a cover. The limitation of manual lubrication is that the chances of contamination are more, sometimes even due to negligence of the worker.

In modern equipment, automatic lubrication systems are provided wherever feasible. Such a system protects the lubricant from adverse atmospheric conditions. There are different types of systems being used such as waste pad, sight feed oilers, lubricators, hydrostatic lubricators, splash oilers, etc. The more recent systems can be classified into:

- Centralized lubrication system (manual)
- Centralized lubrication system (Power operated)

In a manually operated lubrication system, the lubricator discharges lubricants through a main tube to feed the branch tubes connected at tube points. Whereas, in the case of power operated lubrication system, the lubricant is pumped by its own pump, to several lubricating points, on a time-controlled cycle. This system has been very successful and is employed today in many heavy-duty machines that are used in mines. The main advantages of this system are:

- Positive lubrication,
- Reduction in idle time of the machine,
- Reduction in the number of workers,
- Lower maintenance cost.

In the power operated lubrication system, the chances of failure of equipment are reduced, which is possible in the case of manual lubrication system due to neglect of the operator, that is, not providing lubrication in time. But at the same time, failure of the power operated lubrication system can be very serious, since it may seize the operating machine/system.

In some of the cases, it is practically not possible to provide lubricants to the moving parts. Sometimes the conventional methods of lubrication cannot be used. In such cases the sliding surfaces are made of materials which themselves provide some lubrication in the form of process fluid. Sometimes, the moving parts have adequate wear resistance, which can be achieved by applying specially treated coatings. The bearings and seals, operated at very high temperatures, high-pressure, and high speeds, fall under this category. The coatings and surface treatments are applied on the top foil and journal surface to minimize wear of the rubbing surfaces. The selection of suitable coatings depends upon the working environment of the equipment.

SELECTION OF LUBRICANTS

It is the normal practice on the part of the manufacturers to recommend the type of lubricants that are most suitable for their equipment. Such recommended lubricants must always be used. It is a good maintenance practice to prepare a chart of lubricants for important components recommended by the manufacture and display it, near the concerned machine. The proper use of the lubricants pays in the long run by way of increased life and less breakdowns of equipment/parts. Therefore, their selection is of immense importance. The following factors must be considered while selecting a lubricant.

- Type of lubricant i.e. solid or liquid,
- Place of use, for temperature considerations,
- Environmental conditions, i.e. open to atmosphere or sealed,
- Frequency, i.e. at what time intervals the lubricant must be applied, and
- The life of the lubricant.

Planning of Lubrication Schedule

Lack of proper attention towards lubrication of the parts/equipment often leads to a serious problem such as seizure of an engine part. To obviate such adverse effects, it is important that proper thought to lubrication be paid in the following manner:

- Prepare a master chart for lubrication, of every equipment, item-wise and also the due date for replacement of lubricant.
- Workers must be assigned the task of lubrication to complete the work as per schedule.
- Record of every activity relating to the lubrication system must be maintained so that exigencies may be overcome when a particular operator does not attend the duties.

The record of lubrication can be maintained as follows:

(*a*) Records of recommended lubricants and their annual requirement.
(*b*) Consumption of each lubricant equipment-wise.
(*c*) Test records of lubricants for their validity.

(*d*) Type of lubricating system used in each equipment.
(*e*) The persons dealing with the lubrication system must be properly trained.
(*f*) New developments in lubricants and lubricant systems must be followed in order to minimize the maintenance cost.

Handling of Lubricants

In a large organization several types of equipment are normally used which require a variety of lubricants. Contamination will affect the quality of the lubricant after it is produced. Due care should therefore be taken in its package and transportation. It should be packed in clean and airtight containers. The data of manufacturer should also be mentioned on the container to avoid confusion during its use. Lubricants must be safeguarded from getting contaminated due to atmospheric effects, with passage of time. The basic objective should be to maintain its quality.

Handling of lubricants is very important as improper handling of containers can cause serious problem such as fire hazards. Leakage of lubricants from containers will not only cause loss, but also make the place dirty and slippery. The lubricants should not be exposed to high temperatures as that may deteriorate their useful properties and make them unsafe for use.

The conditions, which make lubricants unsuitable for use, are: dirt, water, mixing up with other lubricants, high temperatures as well as low temperatures. Some of the lubricants have specified self-life and their storage for longer periods may thus spoil them. Therefore, it is necessary to adopt an efficient storing system for lubricants and also use them in time.

During transportation of lubricants, due care should be taken so that the containers do not get damaged. For this, single tier storage system must normally be followed. Unloading of the drums or containers should also be handled properly to avoid any damage. Sometimes wooden platforms are used to avoid damage. The same practice should be followed for handling the lubricants in warehouses.

Storing Lubricants

It should be a normal practice to keep the lubricants in a shaded area so that they are not exposed to direct sunlight or rain. Labels must be put on all the containers or tins, so that due care can be taken at the time of issuing the lubricants. Storing the lubricants outside in the open may result in their leakage and loss due to varying temperature. This might cause seals of containers becoming loose and thus enhancing the possibility of contamination due to rain and dust, etc. The cold weather may also adversely affect some soluble oil and compound oils when stored in an open space.

As a temporary measure, the lubricants can be stored outside under a shaded area or under tarpaulin covers. The floor area where the lubricants are stored must be well maintained and the drums must be placed on wooden platforms to keep them off the ground. The drums should never be stored on the ground directly, because this may cause corrosion to metallic parts and thus spoil the lubricants over course of time. Regular inspection of the store must be carried out to check for any damage or leakage of the containers and the identification marks also must be checked regularly.

Drums of different types of lubricants must be stored separately and suitable labels must be affixed to each drum to indicate the type of lubricant stored. Normally, the single-tier system is adopted for storing drums, otherwise wooden planks must be provided in between two drums. While handling the drums, metallic rods or any other sharp objects, which may cause damage to the containers, must be avoided.

In indoor storage of lubricants, due care should be exercised to maintain a fixed temperature inside the warehouse during all the seasons. Precautions should also be taken to avoid fire hazards. Provision must be made to implement the first in first out concept for the stored lubricants in order to avoid contamination due to environmental conditions.

Sealed drums can be stacked horizontally on wooden planks. Forklift may prove better for handling the drums/containers. The oil house must be kept in the cleanest possible state.

The size of the oil house will depend upon the consumption of the lubricants in a plant. Sometimes more than one oil house may be required for storing the lubricants. Entry of the persons to the oil house must be restricted to avoid confusion.

While using drums for storing lubricants, following precautions must be taken:

(*i*) The drums must be placed on a concrete slab and oil is withdrawn directly into measured oil containers for issue to maintenance personnel.

(*ii*) The equipment used for lubricating purposes should be kept in oil racks so that their shape does not deteriorate when not in use; strict care must be exercised to keep the oil room and other dispensing equipment as clean as possible.

(*iii*) For draining oil from the drum corks must be provided and kept in closed position when not in use. They should be checked for any leakage from time to time.

(*iv*) Small trays must be kept underneath while draining oil from drums or to collect the oil drip, if any. Oil thus collected can be filtered and used.

(*v*) Grease drums/containers can be stored vertically and care should always be taken to keep them covered when the lubricant is not being withdrawn from them.

(*vi*) While draining lubricants from bigger containers, the normal practice should be to use measured containers for ensuring the correct amount of the product. The containers used for carrying lubricants from the store should be provided with lids/covers to avoid contamination of lubricants from dust and moisture during transit.

(*vii*) For drawing out grease from the drums, the use of hands should be avoided as far as possible in order to reduce the chances of contamination. Due care should be taken, not to mix up the different grades of grease by any means, which otherwise will affect the performance of the equipment.

(*viii*) For different types of lubricants, colour code schemes must be used to avoid confusion and trained persons must be employed for the job.

(*ix*) In the storeroom, tags indicating the type and grade of lubricant must be displayed for easy identification.

Coding Scheme

To avoid confusion while using lubricants, a colour code scheme should be followed for different types of grades of oils and greases. This will allow easy detection of a particular type of lubricant and also facilitate its storing. Various schemes of symbols can be used such as circle or triangle of different colours for different types of lubricants.

MAINTENANCE OF LUBRICANTS

As stated earlier, to avoid contamination of lubricants it is essential to maintain the oil room as clean as possible with temperature control devices. To prevent penetration of moisture, the inside walls of the oil room are painted. The floor of the room should be cemented so that it can be easily cleaned. No loose materials should be allowed in the oil room to avoid any contamination of lubricants.

The equipment used for handling lubricants must also be maintained with utmost care as far as cleanliness is concerned. Dirt will spoil the lubricants, as discussed earlier. It is also essential to maintain a coding scheme for the lubricants. Labels provided on the containers must be checked from time to time and changed if necessary to avoid confusion. For keeping the material and equipment, the places for each must be specified and this should be followed in practice for easy handling of lubricants.

The expiry dates of lubricants must be checked from time to time to control wastage. Such items must be stored separately to give clear indication.

To avoid spilling, proper dispensing equipment must be used which may include drum, pumps for oils and paddles and air operated pumps for greases. Faucets are mostly used for drawing oils from drums of small size, whereas pumps can be used for large size drums.

For dispensing grease, paddles can be used which are made from wood or metal. The use of metal paddles is normally preferred to avoid contamination. But these must be cleaned properly when different grades of greases are to be dispensed from drums. For semi-fluid grease, air operated pumps can be used.

Though the items of handling equipment are very useful, they must be kept in a clean state, otherwise it will aggravate contamination of lubricants.

PURPOSE OF LUBRICANTS

To minimize friction in the mating parts, the use of lubricants is inevitable whether in the form of solid or fluid. It is generally observed that the proper use of lubricants prolongs the life of equipment and also minimizes unpredicted failures. Therefore, due care should be taken to periodically replenish them in equipment. The working conditions also affect the quality of lubricants, so due care should be taken to assess their life as well.

In the systems where greases are used as lubricants on moving parts exposed to atmosphere, abrasive particles with grease or oil increase the wear rate of mating parts. Samples must be drawn, therefore, from the equipment at frequent intervals to check the quality of lubricants. The condition of equipment must also be monitored periodically as part of the maintenance schedule.

Due to environmental factors, chemical reaction takes place in the lubricants when in use on equipment and this may reduce the life of lubricants. The reactants may be in the form of oxygen, sulphur compounds, phosphorous compounds, and chlorine compounds and organo-metallic compounds which may also react with metals. Therefore, proper attention must be paid to take care of such eventualities in actual practice.

QUESTIONS

1. Enlist the significance of maintenance equipment and facilities.
2. Explain different types of workshops used in any maintenance organization.
3. What is meant by central workshop organization? Explain.
4. How the layouts of the workshops affect the maintenance function? Discuss.
5. Discuss the role of stores for maintenance related activities.
6. Discuss the models used for manpower planning.
7. How lubricants are selected? Explain with their related merits and demerits.
8. Give the reasons for properly storing the lubricants. What are the adverse effects of poorly stored lubricants?
9. Why lubricants are required? Give justification for their use.

Bibliography

Agrawal, S.C., *Maintenance Management*, Prabhu Book Service, Gurgoan.

Buffa, L.S., *Modem Production/Operation Management*, Wiley Eastern, New Delhi, 1985.

Collacott, R.A., *Mechanical Fault Diagnosis and Condition Monitoring*, Chapman and Hall, London, New York, 1982.

Colling, W.C., *Maintenance Management*, American Management Association, New York, 1974.

Corder, A.A., *Maintenance Management Techniques*, McGraw-Hill, New York, 1976.

Cunningham, C.E. and Cox, W., *Applied Maintainability Engineering*, A Wiley Interscience Publication, John Wiley and Sons, New York, 1972.

Frank Gradon, *Maintenance Engineering: Organization and Management*, Applied Science Publication Ltd., London, 1973.

Get, K.H. and Bakh, I.C., *Models of Preventive Maintenance*, North Holland, Amsterdam, 1977.

Gupta, A.K., *Reliability Engineering and Terrotechnology*, Macmillan India Ltd, 1996.

Green, A.E. and Broune, A.J., *Reliability Technology*, John Wiley & Sons, New York, 1977.

Gopalakrishnan, P. and Banerji, A.K., *Maintenance and Spare Parts Management*, Prentice-Hall of India, New Delhi, 1991.

Harriss, Ellya, M.J., *Management of Industrial Maintenance*, Butterworths, London, 1978.

Heintzeiman, J.E., *Complete Handbook of Maintenance*, Prentice-Hall Englewood Cliffs (N.J.), 1976.

John,W. Criwele, *Planned Maintenance for Productivity and Energy Conservation*, The Fairmount Press Inc., Lilbum, GA 30247, 1987.

Kumar, K., *Reliability Modeling and Performance Analysis of Power Distribution Systems for Equipment in Coal Mines*, Ph.D., Thesis, I.S.M. Dhanbad, 1991.

Lindley, R. Higgins, *Maintenance Engineering Handbook*, McGraw Hill Inc.

Mishra, R.C. et al., *'Replacement of Deteriorating Equipment'*, J Institution of Engineers India, Pt 3, vol. 63, Nov. 1982.

Mishra, R.C. and Pathak, K. *Maintenance Engineering and Management*, Prentice-Hall of India Pvt. Ltd., New Delhi, 2002.

Sarker, B.R., *'Manpower Estimation in Continuous Production System—An Analytical Approach'*, J Institution of Engineers India, Pt-ME 3, vol. 63, Nov. 1982.

Shooman, M.L., *Probabilistic Reliability—Engineering Approach*, McGraw-Hill Book Company, New York, 1968.

Srinath, L.S., *Mechanical Reliability*, Affiliated East-West Press Pvt. Ltd., 2002.

Ushakov, I.A., *Reliability Handbook.*

White, E.N., *Maintenance Planning Control and Documentation*, Gower Press Teakfield Ltd., England, 1973.

Williams, J.H., Davies, A. and Drake, *Condition Based Maintenance and Machine Diagnostics*, Chapman & Hall, London, 1994.

Index